Acid-Base Disorders

Basic Concepts and
Clinical Management

Acid-Base Disorders

**Basic Concepts and
Clinical Management**

Heinz Valtin, M.D.   Andrew C. Vail Professor and Chair
of Physiology, Dartmouth Medical School,
Hanover, New Hampshire

F. John Gennari, M.D.   Professor of Medicine and Director, Nephrology
Unit, The University of Vermont College of Medicine,
Burlington, Vermont

Little, Brown and Company : Boston / Toronto

The cover design was inspired by the nomogram for normal human blood taken from L. J. Henderson's classic contribution, *Blood: A Study in General Physiology.* New Haven: Yale University Press, 1928.

To our families

Preface

This book derived its impetus from the concurrence of a fact with a conviction. The fact is the gratifying praise that has been accorded one of us (H.V.) for the chapters on acid-base balance in his two books, *Renal Function* and *Renal Dysfunction.* The conviction concerns the desirability of presenting, in a single volume, the basic principles of acid-base balance and the application of those principles to the analysis and management of clinical problems.

In the first three chapters of the present book we present the physiology of acid-base balance and the practical approach to patients with disturbances of such balance; these chapters resulted from our reworking parallel chapters already published in *Renal Function* and *Renal Dysfunction.* The next four chapters contain, in order, clinical histories of the primary disturbances of acid-base balance: metabolic acidosis, metabolic alkalosis, respiratory acidosis, and respiratory alkalosis. The final chapter deals with mixed disturbances. Major examples for each type of disturbance, drawn largely from the experience of the other of us (F.J.G.), are presented; and in each presentation we take up, successively, the history and physical examination, the laboratory data, the analysis of the problem, and the treatment.

The references at the end of each chapter are meant to serve two major purposes: (1) to document the text (even though specific citations have not been included in the narrative), and (2) to supply further reading for those who wish to pursue a given question in greater detail.

A request for critique. We welcome suggestions from any reader on how this book might be improved.

H. V.
F. J. G.

Acknowledgments

In drawing up a list of those persons who have been of special help with this book, we live in fear of having forgotten someone. Only that individual will spot the inadvertent oversight, and she or he will do us a great favor in bringing the omission to our attention, so we can correct it.

If this book is largely free of error, that is due, in part, to the opportunities afforded us to consult freely with colleagues. Among these, we wish to thank especially: Eugene E. Nattie, Donald Bartlett, Jr., Bruce A. Stanton, and S. Marsh Tenney.

We thank many authors and publishers for allowing us to reproduce their work. Specific acknowledgments are given in the legends to the illustrations.

Finally, we are grateful to the following: Donna C. Bryan, Terry T. Hall, Mary L. Kenyon, and Nancy Perrine for secretarial work; Linda Hare for imaginative help in designing the cover; Joan E. Thomson, Sally D. Whitlock, and Bruce C. Farrell for many of the illustrations; and Curtis R. Vouwie, Clifton Gaskill, and Katharine Tsioulcas at the Medical Division of Little, Brown & Company for their always considerate and sensitive advice and assistance in the various phases of production.

H. V.
F. J. G.

Contents

Acid-Base Disorders

Basic Concepts and
Clinical Management

1 : Acid-Base Balance in Health

The Problem of Mammalian H^+ Balance

The pH of the blood is normally maintained within the small alkaline range of about 7.37 to 7.42. A narrow range of pH is essential to normal metabolic function, probably because the activities of protein macromolecules such as enzymes and of elements required for blood clotting and muscle contraction are importantly influenced by pH. The extreme range of plasma pH that is compatible with life is approximately 6.8 to 7.8.

The pH of the body fluids is maintained alkaline even though the mammalian body normally produces large amounts of acid, from two major sources: (1) The volatile acid H_2CO_3 (see below) is produced from CO_2, the end product of oxidative metabolism, and (2) a variety of nonvolatile acids are produced from dietary substances. As outlined next, the first source is by far the larger.

Some 13,000 to 20,000 mmoles of CO_2 are produced daily as the result of oxidative metabolism; when processed, this CO_2 yields H^+ according to either or both of the following reactions:

$$CO_2 + H_2O \rightleftharpoons H_2CO_3 \rightleftharpoons H^+ + HCO_3^- \qquad (1\text{-}1)$$

and/or

$$\begin{array}{c} HOH \rightleftharpoons H^+ \\ \updownarrow \quad {}_{\text{C.A.}} \\ CO_2 + OH^- \rightleftharpoons HCO_3^- \end{array} \qquad (1\text{-}2)$$

Note that the final products are H^+ and HCO_3^-, whether the hydration of CO_2 is involved (Eq.1-1) or the splitting of H_2O and subsequent hydroxylation of CO_2 (Eq. 1-2). There is evidence that carbonic anhydrase (C.A.) catalyzes reaction 1-2 but not 1-1. Because when either of these reactions moves to the left, the CO_2 formed is rapidly dissipated by the lungs, H_2CO_3 is referred to as a *volatile acid*.

In most western countries, where meat constitutes a large part of the diet, there is also a net daily production of some 40 to 60 mmoles of inorganic and organic acids that are not derived from CO_2. Sulfuric acid is produced from protein catabolism through the conversion of sulfur in the amino acid residues, cysteine, cystine, and methionine, as exemplified by the following reaction for methionine:

$$2\,C_5H_{11}NO_2S + 15\,O_2 \rightarrow 4\,H^+ + 2\,SO_4^{2-} + CO(NH_2)_2 + 7\,H_2O + 9\,CO_2 \qquad (1\text{-}3)$$
$$\text{methionine} \qquad\qquad\qquad\qquad \text{urea}$$

Formation of phosphoric acid during the catabolism of phospholipids makes a minor contribution, as does partial metabolism of carbohydrates and fats, which yields a variety of organic acids. Because all these acids, unlike CO_2, are not volatile or in equilibrium with a volatile component, they are known as *nonvolatile*, or *fixed*, acids. Finally, normal digestive processes result in the loss of some 20 to 40 mmoles of alkali in the stool. In net balance this loss is equivalent to the addition of nonvolatile acid to the body. (When the diet consists mainly of vegetables and fruits, the net production of nonvolatile constituents consists of alkalis.)

In certain physiological and pathological states, the production of nonvolatile acids may rise as much as tenfold. Examples include the production of lactic acid during muscular exercise and states of hypoxia, and the production of acetoacetic acid and 3-hydroxybutyric (β-OH butyric) acid during uncontrolled diabetes mellitus.

Thus, the problem of H^+ balance in most mammals is the defense of normal alkalinity in the face of a constant onslaught of acid.

pH versus [H$^+$]

In recent years a number of experts have advocated the substitution of the hydrogen ion concentration, $[H^+]$, for pH. One can switch from one system to the other through the expression:

$$pH \simeq -\log[H^+] \qquad (1\text{-}4)$$

Some of the major arguments in favor of the use of $[H^+]$ are as follows: (1) It is difficult to handle logarithms, let alone negative logarithms, without the use of log tables or calculators, which are cumbersome tools at the bedside. (2) Concentrations of all substances in clinical medicine, *except H$^+$*, are expressed as quantities per unit volume, such as milligrams or millimoles per liter. (3) In analyzing the movement of substances across membranes, we commonly consider concentration differences or gradients, which are not immediately apparent from a difference in pH. (4) Because the pH scale is logarithmic, an equal change in pH may reflect unequal changes in H^+ concentration; thus, a change in pH from 7.0 to 7.1 represents a change in $[H^+]$ from

100 to 79 nanomoles (nmoles) per liter, whereas a change in pH from 7.3 to 7.4 represents only one-half as great a change in $[H^+]$, namely, from 50 to 40 nmols per liter.

Perhaps equally compelling arguments are marshaled by those who advocate the continued use of pH: (1) Unlike that of most other substances in clinical medicine, the concentration of H^+ cannot be measured directly, but is in fact estimated by determining the pH. (2) The modern laboratory now routinely measures or derives — and reports — all three variables of the Henderson equation (Eq. 3-4) and of the Henderson-Hasselbalch equation (Eq. 1-9); this practice eliminates the need for computations involving logarithms at the bedside. (3) Like that of other substances, the biological activity of H^+ is a function of its chemical potential, which is much more closely related to the logarithm of the $[H^+]$ than to the $[H^+]$ itself; therefore, equal changes in pH may more nearly reflect equivalent physiological effects of H^+ than would equal changes in $[H^+]$.

As we shall see in Chapter 3, most experts on acid-base balance now utilize both pH and $[H^+]$ in managing patients, and we shall follow that practice in this book.

Buffering of Nonvolatile (Fixed) Acids

The very effective defense of alkalinity in a normal dog is illustrated in Figure 1-1. It compares the change in the pH of arterial plasma when 156 ml of a 1 N HCl solution was infused intravenously with the drop in pH when the same amount of acid was gradually added to 11.4 liters of distilled water. This volume of distilled water is about equal to the total body water of the dog.

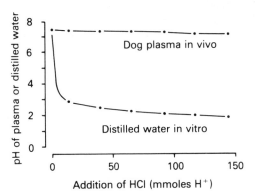

Fig. 1-1 : An experiment contrasting the effective buffering of HCl in a dog with the lack of buffering when the same amount of acid is added to distilled water. The pH of the dog's arterial plasma decreased gradually from 7.44 to 7.14; that of unbuffered distilled water dropped precipitously to a level that would be fatal if it occurred in vivo. Redrawn from Pitts, R. F., *Harvey Lect.* 48:172, 1953.

In the dog the pH dropped from 7.44 to 7.14, a state of severe acidosis but one compatible with survival. In contrast, the addition of just a few millimoles of H^+ to unbuffered distilled water sharply lowered the pH to a value that would have been fatal to the animal, and the final level was 1.84. This section deals with the mechanisms that permit such effective buffering in vivo.

First Line of Defense — Fast Physicochemical Buffering

The following reaction is the prototype for physicochemical buffering:

Strong acid + Buffer salt $\rightleftharpoons$ Neutral salt + Weak acid   (1-5)

If hydrochloric acid is buffered by the bicarbonate buffer system, the reaction is:

$$H^+ + Cl^- + Na^+ + HCO_3^- \rightleftharpoons Na^+ + Cl^- + H_2CO_3 \quad (1-6)$$

Insofar as physicochemical buffering reduces the amount of buffer salt and increases the amount of weak acid, this type of reaction only minimizes, but by no means prevents, a decrease in pH. This point can be illustrated by simple calculations that utilize the derivation of the Henderson-Hasselbalch equation as it applies to the bicarbonate system:

$$pH = pK' + \log \frac{[HCO_3^-]}{[H_2CO_3]} \quad (1-7)$$

The pK' (negative logarithm of the apparent dissociation constant) in Equation 1-7 is 3.5. Carbonic acid is in equilibrium with CO_2 (Eq. 1-1), and at the temperature and ionic concentration of the body fluids — and as long as carbonic anhydrase is present — there are approximately 400 molecules of dissolved CO_2 for every molecule of H_2CO_3. For that reason, a physiologically more meaningful form of Equation 1-7 is:

$$pH = pK' + \log \frac{[HCO_3^-]}{[\text{Dissolved } CO_2 + H_2CO_3]} \quad (1-8)$$

In this equation the pK' has a value of 6.1, reflecting the fact that the denominator is now some 400-fold higher than the one in Equation 1-7. In the sense that the denominator of Equation 1-8 consists so overwhelmingly of CO_2 and not of H_2CO_3, the acid moiety of the bicarbonate buffer system is CO_2, even though it cannot donate H^+ (also called protons); in fact, some authorities speak of the "HCO_3^-/CO_2" buffer system rather than of the "HCO_3^-/H_2CO_3" system, and we shall use the former expression from now on.

The concentration of dissolved CO_2 in plasma is proportional to the partial pressure of CO_2 (PCO_2) in the plasma, which is relatively easy to determine. The proportionality constant for plasma at 37°C, which converts PCO_2 in millimeters of mercury (mm Hg) to concentration of dissolved CO_2 expressed as millimoles per liter (mmoles/liter), is 0.03. Thus, if we ignore the trace amounts that exist as H_2CO_3, the denominator in Equation 1-8 can be very closely approximated as $PCO_2 \times 0.03$, and this equation may be rewritten in a form that is most useful in physiological and clinical practice:

$$pH = 6.1 + \log \frac{[HCO_3^-]}{0.03 \times PCO_2} \tag{1-9}$$

In this equation, $[HCO_3^-]$ is expressed as mmoles/liter and PCO_2 in mm Hg. Substituting values for arterial plasma of man in normal H^+ balance (see Table 1-1):

$$pH = 6.1 + \log \frac{24 \text{ mmoles/L}}{0.03 \times 40 \text{ mm Hg}}$$

$$pH = 6.1 + \log \frac{24 \text{ mmoles/L}}{1.2 \text{ mmoles/L}} \tag{1-10}$$

$$pH = 6.1 + \log 20$$

$$pH = 7.40$$

If 12 mmoles of HCl were added to each liter of extracellular fluid — and if, for the moment, we say that all the acid is buffered by HCO_3^- — then physicochemical buffering would decrease the numerator and increase the denominator by 12 mmoles/liter each, according to the following reaction:

$$12\,H^+ + 12\,Cl^- + 24\,Na^+ + 24\,HCO_3^- \rightleftharpoons 12\,Na^+ + 12\,Cl^- + 12\,Na^+ + 12\,HCO_3^-$$
$$+ \; 12\,H_2CO_3$$
$$\Updownarrow$$
$$12\,CO_2 + 12\,H_2O \tag{1-11}$$

If this reaction were to occur in a "closed system" — i.e., without a ventilatory system to eliminate the newly generated CO_2 — the pH would drop to the fatal level of 6.06:

$$pH = 6.1 + \log \frac{12 \text{ mmoles/L}}{1.2 + 12 \text{ mmoles/L}}$$

$$pH = 6.1 + \log \frac{12 \text{ mmoles/L}}{13.2 \text{ mmoles/L}}$$

$$pH = 6.06$$

This dire consequence is prevented by the second line of defense, which, like physicochemical buffering, comes into play within seconds or minutes after the administration of HCl.

Second Line of Defense — Fast Respiratory Component

Because of the equilibrium in Equation 1-1, virtually all the H_2CO_3 that was produced through physicochemical buffering is converted to CO_2 and H_2O (Eq. 1-11), and the CO_2 is excreted by the lungs. If all the extra CO_2 were excreted, returning the denominator to 1.2 mmoles/liter, the resulting pH would fall into the range that is compatible with survival.

$$pH = 6.1 + \log \frac{12 \text{ mmoles/L}}{1.2 \text{ mmoles/L}}$$

$$pH = 6.1 + \log 10$$

$$pH = 7.10$$

Respiratory compensation goes further, however. As a result of the lower pH of the blood, alveolar ventilation is increased, so that alveolar and hence arterial P_{CO_2} are decreased. Consequently the pH is returned toward, but not quite to, the normal value.

$$pH = 6.1 + \log \frac{12 \text{ mmoles/L}}{0.03 \times 23 \text{ mm Hg}}$$

$$pH = 6.1 + \log \frac{12 \text{ mmoles/L}}{0.69 \text{ mmoles/L}}$$

$$pH = 7.34$$

Third Line of Defense — Slow Renal Component

Although respiratory compensation has, within minutes, restored the pH almost to normal, the body stores of the main extracellular buffer have been depleted. This fact is reflected in the decrease of the HCO_3^- concentration from 24 to 12 mmoles/liter. Furthermore, some of the added H^+, although admittedly no longer in free solution, still remains within the body as weak acid. Both of these remaining abnormalities are corrected by the kidneys, which excrete H^+ and simultaneously replenish the depleted HCO_3^- stores. This process is a much slower one than the first two lines of defense, requiring hours to days rather than seconds or minutes. How the kidneys accomplish this task, which finally restores H^+ balance, is discussed in Chapter 2.

The above is a dramatic example that occurs only under artificial experimental conditions or in disease states. Nevertheless, these are the pathways by which the daily loads of nonvolatile acids are handled. The following quantitative comparison may put into perspective the normal daily challenge from these fixed acids. An adult weighing 70 kg has about 14 liters of extracellular fluid (about 20% of body weight). Hence the addition of 12 mmoles of HCl to each liter of extracellular fluid, as in the example described above, would be a total acid load to an adult human of 168 mmoles (12 mmoles/liter · 14 liters). The normal daily load of 40 to 60 mmoles of fixed acids is only about one-third of that amount, and it is released relatively slowly over a 24-hour period, rather than in 1 to 2 hours, as in the above example. If, for the sake of illustration, about one-third of a total load of 48 mmoles were released after each meal, 16 mmoles (48 ÷ 3) would be added to 14 liters of extracellular fluid, i.e., an addition of about 1 mmole per liter of extracellular fluid. Since sulfuric acid is normally the major fixed acid (Eq. 1-3), the quantitative reaction would be as follows:

$$2\,H^+ + SO_4^{2-} + 24\,Na^+ + 24\,HCO_3^- \rightleftharpoons 2\,Na^+ + SO_4^{2-} + 22\,Na^+ + 22\,HCO_3^-$$
$$+ 2\,H_2CO_3$$
$$\updownarrow$$
$$2\,CO_2 + 2\,H_2O \qquad (1\text{-}12)$$

Despite the apparent lowering of $[HCO_3^-]$ from even this small addition of acid, the arterial pH actually does not change. This is so because, in a normal individual in the steady state, the processes described above as the first and second lines of defense are accompanied by the renal excretion of H^+ and the reabsorption of HCO_3^- (the third line of defense, described in Chap. 2).

Buffering of the Volatile "Acid" CO_2

At the beginning of this chapter we cited the fact that some 13,000 to 20,000 mmoles of CO_2 is produced daily by an adult person, as the result of metabolic events. As shown in Equations 1-1 and 1-2, the processing of this CO_2 can generate H^+, and the production of CO_2 therefore potentially disturbs H^+ balance. Ultimately, elimination of all the extra CO_2 by the lungs prevents acidosis (Eq. 1-9); but before that elimination can occur, defense of alkalinity is threatened as the CO_2 is carried in the blood from the cells, where it is produced, to the lungs, where it is excreted. The extremely effective buffering in the blood is reflected in the fact that the difference in pH between venous blood, which goes to the lungs, and arterial blood, which leaves them, seldom exceeds 0.04 of a pH unit. This section deals with the mechanisms by which H^+ is buffered as CO_2 is transported in the blood.

Transport of CO_2 in Blood

The plasma P_{CO_2} at the arterial end of tissue capillaries is about 40 mm Hg. Since the P_{CO_2} is higher in tissue cells that produce CO_2, the gas will diffuse from the tissue cells into the capillary.

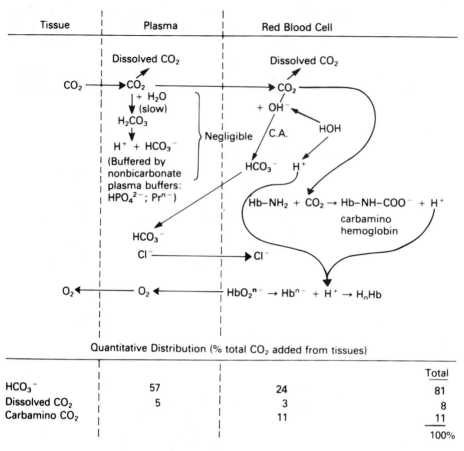

Tissue	Plasma	Red Blood Cell

Quantitative Distribution (% total CO_2 added from tissues)

			Total
HCO_3^-	57	24	81
Dissolved CO_2	5	3	8
Carbamino CO_2		11	11
			100%

Fig. 1-2 : Transport of CO_2 and buffering of H^+ by the blood. Adapted from Davenport, H. W., *The ABC of Acid-Base Chemistry* (6th ed.). Chicago: University of Chicago Press, 1974; and Masoro, E. J., and Siegel, P. D., *Acid-Base Regulation: Its Physiology, Pathophysiology and the Interpretation of Blood-Gas Analysis* (2nd ed.). Philadelphia: Saunders, 1977.

The chain of events that then occurs, shown in Figure 1-2, is as follows:

1. There is still no agreement whether CO_2 is processed mainly by hydration (Eq. 1-1) or by hydroxylation (Eq. 1-2). Because it is the latter reaction that is probably catalyzed by carbonic anhydrase (C.A.), we will use Equation 1-2 to show the processing of CO_2 within cells where C.A. is present, and Equation 1-1 for extracellular fluids where C.A. is absent.

Most cell membranes, including those of red blood cells (erythrocytes), are highly permeable to CO_2. Hence CO_2 diffuses not only into the plasma but also into the erythrocytes. Because there is much carbonic anhydrase in erythrocytes but none in plasma, CO_2 is processed much more rapidly within these cells than in the plasma. In fact, the processing is negligible in plasma; the small amount of H^+ that is formed from this reaction is buffered by the nonbicarbonate buffer anions of plasma, the proteins (Pr^{n-}) and phosphate (HPO_4^{2-}).

2. The rapid combination of CO_2 with OH^- within the erythrocytes yields HCO_3^-. Most of the newly formed HCO_3^- diffuses into the plasma, and Cl^- shifts into the erythrocytes. In this way, most of the CO_2 that is added to venous capillary blood is carried to the lungs as HCO_3^- in the plasma. A portion combines with hemoglobin to form carbamino hemoglobin, and an even smaller amount is carried as dissolved CO_2 within the erythrocytes. The H^+ formed as water is split is buffered, primarily by hemoglobin. The same is true of the H^+ that is released during the formation of carbamino hemoglobin.

3. In a normal, resting adult human, every liter of venous blood that goes to the lungs carries about 1.68 mmoles of extra CO_2 for excretion. The quantitative distribution of the various forms in which the added CO_2 is carried is shown at the bottom of Figure 1-2. About 81% of the 1.68 mmoles is carried as HCO_3^-, most of which is carried in the plasma, even though virtually all of it was generated within the erythrocytes. The remainder is divided between dissolved CO_2 and carbamino CO_2. Of these, the major portion of dissolved CO_2 is carried in the plasma, whereas practically all the carbamino CO_2 is found in the erythrocytes.

Hemoglobin as a Buffer

As CO_2 is added to venous blood, the pH drops from about 7.40 in arterial blood to only about 7.37 in venous blood, rather than to about 7.32, which would be predicted. The mechanisms underlying this effect are shown in Figure 1-3.

The pK of oxygenated hemoglobin (HbO_2^{n-}) is lower than the pK of deoxygenated hemoglobin (Hb^{n-}; see Table 1-2); i.e., Hb^{n-} is less acidic than HbO_2^{n-}. As blood enters the arteriolar end of tissue capillaries, it gives up O_2 to the cells. The consequent reduction of HbO_2^{n-} to Hb^{n-} *would* cause a tremendous rise in pH were it not for the fact that CO_2, and hence H^+, is simultaneously added to the system. The net result of the change in pK of hemoglobin is that roughly 1.3 mmoles of CO_2 can be added to each liter of venous blood as it flows through the tissue capillaries, without changing the pH of that blood. Approximately 92% of the CO_2 that is added to each liter of venous blood, or about 1.6 mmoles of CO_2, is converted to H^+ (Fig. 1-2). Since 1.3 mmoles could be added without a change in pH, only about 0.3 mmole needs to be buffered by Hb^{n-}, and the drop in pH of venous blood is thus minimized.

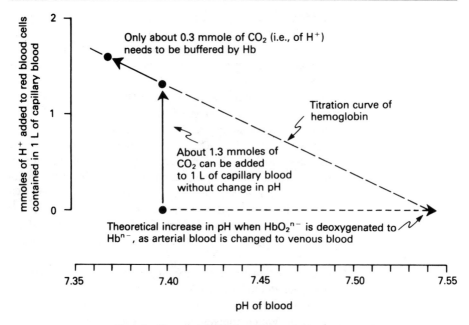

Fig. 1-3 : The special attributes of hemoglobin as a buffer. The pK of deoxygenated hemoglobin (Hb^{n-}) is higher than that of oxygenated hemoglobin (HbO_2^{n-}). Therefore, hemoglobin becomes a more effective buffer at precisely the moment when CO_2 and hence H^+ are added from the tissue cells to the blood.

Concept of Metabolic and Respiratory Disturbances

Primary Disturbances

Inspection of the Henderson-Hasselbalch equation, as represented in Equation 1-9, makes clear that an abnormality of plasma pH can result from a primary deviation of either the $[HCO_3^-]$ or the PCO_2. Since the latter is regulated by the rate of alveolar ventilation, any disturbance in pH that results from a primary change in PCO_2 is called a respiratory acid-base disorder. Hypoventilation and retention of CO_2 lead to a reduction in pH that is called *respiratory acidosis;* hyperventilation and a fall in PCO_2 lead to a rise in pH that is called *respiratory alkalosis.* Changes in the concentration of HCO_3^- are brought about most commonly by the addition or loss of nonvolatile (fixed) acids or bases, which are derived mainly from metabolic processes. Hence, any abnormality of pH resulting from a change in $[HCO_3^-]$ is called a metabolic acid-base disorder. A primary reduction in $[HCO_3^-]$, termed *metabolic acidosis,* can occur when endogenous acids are produced faster than they can be excreted, when HCO_3^- is lost from the body, or when exogenous acid is administered. A primary increase in $[HCO_3^-]$, termed *metabolic alkalosis,* can occur when endogenous HCl is lost from the body or when exogenous HCO_3^- is administered.

Thus, there are four primary disturbances of H^+ balance (often called acid-base balance): (1) respiratory acidosis, (2) respiratory alkalosis, (3) metabolic acidosis, and (4) metabolic alkalosis. Examples of these disorders are discussed in detail in Chapters 4 through 7.

Mixed Disturbances

Not infrequently, two primary disturbances, usually one respiratory and the other metabolic, occur simultaneously in the same individual. Such patients are said to have "mixed" acid-base disturbances. For example, a patient who manifests alveolar hypoventilation from emphysema may also have an obstructed duodenal ulcer leading to loss of HCl through vomiting. This patient would have a mixed disturbance of respiratory acidosis and metabolic alkalosis. Another patient may have both emphysema, with retention of CO_2, and renal failure, which leads to the retention of fixed acids. This patient would have a mixed disturbance of respiratory acidosis and metabolic acidosis. Examples of mixed acid-base disturbances are discussed in detail in Chapter 8.

Compensatory Responses

Primary disturbances in H^+ balance elicit a secondary response that partially corrects the pH. In the example cited earlier in this chapter, the addition of HCl led to a decrease in $[HCO_3^-]$ and hence to metabolic acidosis. This disturbance was largely compensated for by the second line of defense, in which alveolar hyperventilation lowered the P_{CO_2} and thereby adjusted the pH to a near normal value.

This example illustrates two points: (1) that a compensatory response involves the system opposite to the one that caused the primary disturbance (e.g., *metabolic* alkalosis is compensated for by a *respiratory* response, and vice versa) and (2) that compensation shifts the pH *toward* but not to the normal value. Regarding the second point, in the example given earlier in this chapter, respiratory compensation moved the pH to 7.34 but not entirely into the normal range.

Since primary disturbances may be either metabolic or respiratory in origin, and may cause either acidosis or alkalosis, there are four general types of compensatory responses: (1) Metabolic acidosis is compensated for by alveolar hyperventilation. (2) Metabolic alkalosis is compensated for by a reduction in alveolar ventilation. (3) Respiratory acidosis is compensated for by increased renal excretion of H^+ and increased renal reabsorption of HCO_3^- (see Chap. 2 for mechanisms by which this response is accomplished). (4) Respiratory alkalosis is compensated for by decreased renal excretion of H^+ and decreased renal reabsorption of HCO_3^-.

Compensations for primary metabolic disturbances occur virtually immediately, whereas those for primary respiratory disturbances are fully manifested only after several days. (This fact is discussed further in Chap. 3, under Confidence Bands and Rules of Thumb, as well as at the beginning of Chap. 6.) The reason is that changes in alveolar ventilation almost instantly alter the level of PCO_2, while changes in renal function that alter the plasma $[HCO_3^-]$ take much longer. It is because of this difference in the time course of compensatory responses that primary respiratory disturbances — but not primary metabolic disturbances — are subdivided into acute (up to 8 hours) and chronic (present for 2 to 3 days or longer) phases. Between these time intervals a transitional phase is present, reflecting the gradual development of the renal response. During acute respiratory acidosis or alkalosis there will be a very slight rise or fall, respectively, of the plasma $[HCO_3^-]$ resulting from the chemical reactions shown in Equations 1-1 and 1-2, as well as from buffering by nonbicarbonate buffers of the H^+ produced in these reactions (Eqs. 1-15 and 3-7; see also the beginning of Chap. 6). In chronic respiratory disturbances the changes in plasma $[HCO_3^-]$ are much greater because the kidneys have compensated by altering the reabsorptive rate for HCO_3^- (described further in Chap. 2).

The Important Buffers of Mammals

Thus far we have spoken almost exclusively about the bicarbonate and the hemoglobin buffer systems. In this section we shall emphasize that the body contains other important buffers, and that all these participate in the regulation of pH.

A buffer is a mixture either of a weak acid and its conjugate base or of a weak base and its conjugate acid. (Acid is here defined as a H^+ donor and base as a H^+ acceptor.) The buffers of physiological importance in mammals are all of the first type. They have been listed in Figure 1-4. These buffer systems are by no means limited to the plasma. They are found in all phases of the body fluids — plasma, interstitial fluid, intracellular fluid — and bone. The bicarbonate system predominates in plasma and interstitial fluid, while organic phosphates and proteins (especially hemoglobin) predominate in the intracellular space.

The Isohydric Principle

When several buffers exist in a common solution, as in a beaker, all the buffer pairs are in equilibrium with the same concentration of H^+. Expressed in the terminology of the Henderson-Hasselbalch equation, and for plasma — which is a common

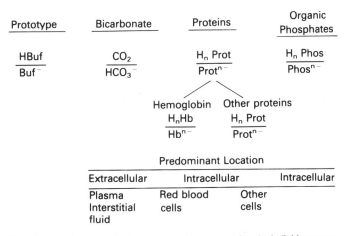

Prototype	Bicarbonate	Proteins	Organic Phosphates
$\dfrac{HBuf}{Buf^-}$	$\dfrac{CO_2}{HCO_3^-}$	$\dfrac{H_n\,Prot}{Prot^{n-}}$	$\dfrac{H_n\,Phos}{Phos^{n-}}$

Hemoglobin Other proteins
$\dfrac{H_nHb}{Hb^{n-}}$ $\dfrac{H_n\,Prot}{Prot^{n-}}$

Predominant Location

Extracellular	Intracellular	Intracellular
Plasma Interstitial fluid	Red blood cells	Other cells

Fig. 1-4 : The important buffer systems of the mammalian body fluid compartments. Note that CO_2 is denoted as the acid moiety of the bicarbonate buffer system, even though CO_2 is not a H^+ donor (see text discussion of Eq. 1-8). The locations refer to quantitative predominance and are not exclusive except for hemoglobin in erythrocytes. The valence of phosphate is designated as indefinite because it is quantitatively an important chemical buffer mainly within the intracellular fluid, where its valence as organic phosphate is not known.

solution containing the bicarbonate, protein, and phosphate buffer systems — the isohydric principle can be stated as follows:

$$pH = pK'_1 + \log \frac{[HCO_3^-]}{[CO_2]} = pK'_2 + \log \frac{[HPO_4^{2-}]}{[H_2PO_4^-]} = pK'_3 + \log \frac{[Prot^{n-}]}{[H_nProt]} \quad (1\text{-}13)$$

This principle has an important application in the analysis of acid-base disturbances, because one can infer the status of most of the body buffer pairs by determining the status of just one of them. In practice, one usually measures two of the three variables of the bicarbonate system, so that the third can easily be calculated (Eq. 1-9). For plasma, which is a homogeneous solution, knowledge of the bicarbonate system can thus be extended precisely to the phosphate and protein buffer pairs without actually measuring their concentrations.

Precise extension to the buffers in the other major fluid compartments is not possible because these compartments are not part of a homogeneous solution, in which all the buffers are evenly distributed. For example, the pH of the intracellular compartment is considerably lower (perhaps 7.00) than that of the extracellular compartment, because of three influences: (1) metabolic production of acids within cells, which is modified by (2) active transport of H^+ out of cells, and (3) the so-called Gibbs-Donnan effect, which results in a slightly unequal distribution of diffusible electrolytes on two sides of a semipermeable membrane. The Gibbs-Donnan effect accounts for the approximately 5% difference in buffer concentrations between the plasma and interstitial fluid. Despite these differences among compartments, however, the isohydric principle can be applied in many conditions of H^+ imbalance to infer qualitative changes in all or most of the body buffers from knowledge of the bicarbonate system alone. This is permissible because many acid-base disturbances represent relatively chronic situations in which the change in H^+ balance within one fluid compartment has been accompanied by qualitatively similar changes in the other compartments.

Special Attributes of the Bicarbonate System

We have already reviewed the special properties of hemoglobin that render it extraordinarily suitable for buffering the H^+ that is formed when CO_2 is added to venous blood. In turn, the bicarbonate system has special attributes for buffering nonvolatile acids. Titration curves for the bicarbonate and inorganic phosphate buffers, which are found in plasma and interstitial fluid, are shown in Figure 1-5. Several points should be noted: (1) the pK' is numerically equal to the pH existing when the weak acid and its conjugate base each compose 50% of the total concentration of that particular buffer; (2) the change in pH per quantum of H^+ or OH^- added is least in the linear portion of each titration curve; and (3) this linear portion of most effective chemical buffering extends roughly 1.0 pH unit to either side of the pK' — from pH of about 5.1 to 7.1 for the bicarbonate system, and from about 5.8 to 7.8 for phosphate. In other words, in the range of

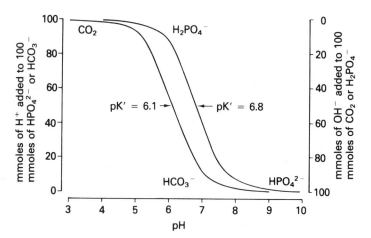

Fig. 1-5 : Titration curves for the bicarbonate and inorganic phosphate buffers in a closed system. Under these conditions, when the concentration of CO_2 cannot be kept low through diffusion to the outside — and at the pH of body fluids — the bicarbonate system is a less effective buffer than phosphate. It is mainly because CO_2 can normally be eliminated by the lungs that the bicarbonate system is such an efficient physiological buffer. Modified from Pitts, R. F., *Physiology of the Kidney and Body Fluids* (3rd ed.). Chicago: Year Book, 1974.

plasma pH that is compatible with survival (about 6.8 to 7.8), each quantum of phosphate can buffer more H^+ than an equal quantum of bicarbonate. Nevertheless, bicarbonate plays a much more important physiological role as an extracellular buffer than does phosphate.

The reason for the seeming paradox is not only that the extracellular concentration of bicarbonate (about 24 mmoles/liter) is so much higher than that of phosphate (1 to 2 mmoles/liter; see Table 1-1) but also, and more importantly, that bicarbonate has certain physiological properties that make it a uniquely effective buffer. The central property is that carbonic acid is in equilibrium with volatile CO_2, which can be rapidly excreted or retained by the lungs (see Second Line of Defense, earlier in this chapter). Furthermore, both the acid moiety, CO_2, and the conjugate base, HCO_3^-, of the bicarbonate buffer system are more abundantly available from daily metabolic processes than are those of any other buffer system.

Utilization of the Various Buffers

Addition of Strong Acid

Earlier in this chapter, when we described the buffering of non-volatile (fixed) acids — i.e., of H^+ other than that generated by CO_2 (Eqs. 1-1 and 1-2) — we limited the analysis to the bicarbonate buffer system in plasma. This simplification was not seriously inaccurate, since an acid that is infused intravenously will initially have its impact on the plasma. The description, however, did not include the participation of the other buffer systems of the body. The total picture is presented in Figure 1-6, which indicates not only the time course for distribution of a fixed acid throughout the body fluid compartments but also the quantitative contribution of the major buffers. (The figure refers to the addition of an inorganic, or mineral, acid; details may be slightly different when the same amount of H^+ is added in the form of an organic acid, such as lactic acid.)

Figure 1-6 is based on experimental work in dogs that were given intravenous infusions of hydrochloric acid over periods of 1.5 to 3 hours. Within seconds to minutes, the acid was being handled by the various buffers of the blood. In the plasma, this process involves mainly the bicarbonate system, because its ionic concentration in plasma is so much greater than that of the proteins and inorganic phosphate (Table 1-1), and because CO_2, which is formed in the buffering reaction, is quickly eliminated via the lungs. The acid also quickly enters the erythrocytes, where it is buffered primarily by hemoglobin; bicarbonate, and to an even lesser extent organic phosphates, within erythrocytes contribute a little bit to the buffering. A small amount of the acid

16

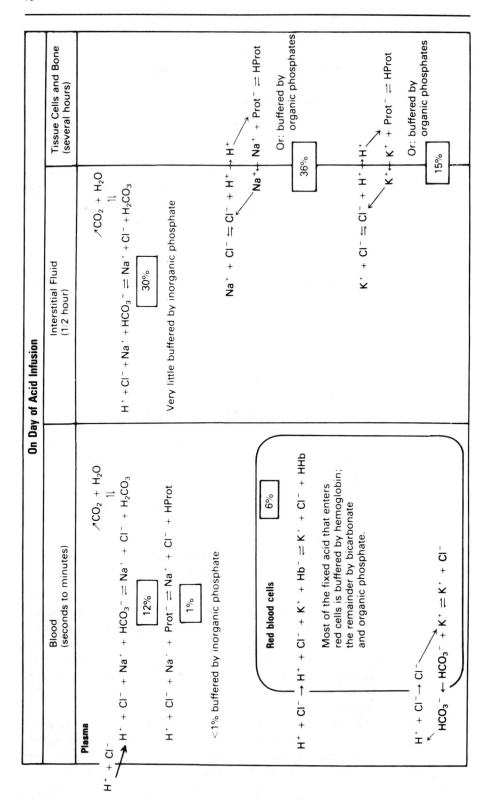

On Day of Acid Infusion

	Blood (seconds to minutes)	Interstitial Fluid (1/2 hour)	Tissue Cells and Bone (several hours)

$H^+ + Cl^-$

Plasma

$\nearrow CO_2 + H_2O$
$\updownarrow$
$H^+ + Cl^- + Na^+ + HCO_3^- \rightleftharpoons Na^+ + Cl^- + H_2CO_3$

$\boxed{12\%}$

$H^+ + Cl^- + Na^+ + Prot^- \rightleftharpoons Na^+ + Cl^- + HProt$

$\boxed{1\%}$

<1% buffered by inorganic phosphate

Red blood cells $\boxed{6\%}$

$H^+ + Cl^- \rightarrow H^+ + Cl^- + K^+ + Hb^- \rightleftharpoons K^+ + Cl^- + HHb$

Most of the fixed acid that enters red cells is buffered by hemoglobin; the remainder by bicarbonate and organic phosphate.

$H^+ + Cl^- \rightarrow Cl^-$
$HCO_3^- \leftarrow HCO_3^- + K^+ \rightleftharpoons K^+ + Cl^-$

$\nearrow CO_2 + H_2O$
$\updownarrow$
$H^+ + Cl^- + Na^+ + HCO_3^- \rightleftharpoons Na^+ + Cl^- + H_2CO_3$

$\boxed{30\%}$

Very little buffered by inorganic phosphate

$Na^+ + Cl^- \rightleftharpoons Cl^- + H^+ \rightarrow H^+$

$Na^+ \rightarrow Na^+ + Prot^- \rightleftharpoons HProt$

$\boxed{36\%}$

Or: buffered by organic phosphates

$K^+ + Cl^- \rightleftharpoons Cl^- + H^+ \rightarrow H^+$

$K^+ \rightarrow K^+ + Prot^- \rightleftharpoons HProt$

$\boxed{15\%}$

Or: buffered by organic phosphates

24 Hours After Infusing Acid

About 25% of acid load has been excreted in the urine as titratable acid and NH_4^+ (see Chap. 2). Extracellular pH and ionic composition are nearly normal; therefore, 75% of the administered acid must be sequestered and buffered in tissue cells and bone.

2-6 Days After Infusing Acid

The remaining 75% of the administered acid is slowly released from tissue cells and bone, and excreted in the urine.

Fig. 1-6 : Handling of the fixed inorganic acid, HCl, by intact dogs. For the sake of clarity, polyvalent anions such as proteins and hemoglobin have been drawn with a single negative sign. A slanted arrow next to CO_2 indicates that the CO_2 is quickly excreted through the lungs. The percentages enclosed in rectangles indicate the approximate proportion of the total acid load that is buffered by each mechanism. Data from Swan, R. C., and Pitts, R. F., J. Clin. Invest. 34:205, 1955; Yoshimura, H., et al., Jpn. J. Physiol. 11:109, 1961; Pitts, R. F., Physiology of the Kidney and Body Fluids (3rd ed.). Chicago: Year Book, 1974; and Masoro, E. J., and Siegel, P. D., Acid-Base Regulation: Its Physiology, Pathophysiology and the Interpretation of Blood-Gas Analysis (2nd ed.). Philadelphia: Saunders, 1977.

is buffered in the plasma, by bicarbonate that was derived from the erythrocytes through an exchange of Cl^- for HCO_3^-.

As soon as the HCl is infused into the plasma, it begins to enter the interstitial compartment. Although it takes about one-half hour for the acid to be evenly distributed between the plasma and interstitial fluid, the latter actually contributes more to the total buffering because its volume is about four times greater than that of plasma. Again, inorganic phosphate in the interstitium also participates, but to a negligible extent, because its concentration in interstitial fluid, as in plasma, is very low.

Several hours elapse before the acid is evenly distributed throughout the intracellular compartment. Nevertheless, the contribution of the intracellular buffers is great. Utilization of these buffers — mainly proteins and organic phosphates — is accomplished by exchange of extracellular H^+ for either intracellular Na^+ or intracellular K^+. Some of the Na^+ probably comes from the apatite of bone, and the H^+ that is exchanged for this Na^+ enters into a chemical reaction with the apatite, finally being incorporated into HCO_3^-. The involvement of bone in buffering is probably much more important during chronic disturbances of acid-base balance (as in chronic renal failure) than during the relatively acute disturbance illustrated in Figure 1-6.

The relative quantitative contributions of the various buffers are indicated by the boxes in Figure 1-6, which give the percentage of the total acid load that is handled by each mechanism. It is clear that less than 15% is buffered in the plasma, and less than 20% by whole blood. To be emphasized is the fact that about one-half of the administered acid is buffered in the intracellular compartment.

Although buffering of the acid has minimized changes in pH, restoration of external balance must await excretion of the acid load in the urine. This occurs during the following several days. By the second day after the infusion, about 25% of the acid has been excreted, and both the pH and ionic composition of the extracellular fluids have returned to near normal values. It follows that 75% of the acid load must now reside within the cells and bone, where it is buffered. This remaining acid is slowly released into the extracellular fluids and is excreted by the kidneys during the second to sixth days after the acid was given.

Addition of Strong Alkali

When base is infused intravenously, or HCl is eliminated, as by vomiting, the chain of events is similar to that discussed above, except that the reactions go in the opposite direction. There is a marked participation by the intracellular buffers, and the time course for the utilization of the various buffers is similar to that shown in Figure 1-6. Three differences should be noted, however: (1) respiratory compensation — i.e., alveolar hypoventilation due to an increase in pH — is less intense in metabolic alkalosis than in metabolic acidosis (Table 3-3); (2) lactic acid moves out of skeletal muscle cells, to buffer the base in the extracellular fluid; and (3) the renal excretion of base, such as sodium bicarbonate, usually occurs more rapidly than does the renal excretion of H^+. The last two effects occur only when the alkalosis is due to addition of base rather than to loss of acid.

Addition of the Volatile "Acid" CO_2 — Hypercapnia

If excess CO_2 is produced endogenously and there is no disorder of respiration, the surplus is quickly excreted through the lungs by the mechanisms shown in Figure 1-2. If, however, CO_2 is added from an external source, as by breathing a gas mixture containing 5% CO_2, or if CO_2 accumulates because of some disorder of respiration, there is a net addition of acid to the body by the reactions shown in Equations 1-1 and 1-2. The chain of events that is set into motion under these circumstances is shown in Figure 1-7.

The first important point to recognize is that the H^+ produced when CO_2 is added cannot be buffered by the bicarbonate system. The reason is evident from Equation 1-14. When H^+ is buffered by HCO_3^- (as in Eqs. 1-11 and 1-12), carbonic acid is formed momentarily, and the final products are CO_2 and H_2O:

$$H^+ + HCO_3^- \rightleftharpoons H_2CO_3 \rightleftharpoons CO_2 + H_2O \qquad (1\text{-}14)$$

Since CO_2 and H_2O are the starting substrates when CO_2 is added to the body (Eqs. 1-1 and 1-2), reaction 1-14, under that circumstance, is being driven to the left, and it cannot simultaneously be driven to the right, as would be required if the H^+ were to be buffered by HCO_3^-. Instead, the H^+ must be buffered by the nonbicarbonate buffers (proteins and phosphates), designated by Buf^- in the following reaction:

$$CO_2 + OH^- \overset{C.A.}{\rightleftharpoons} HCO_3^- \rightarrow HBuf + Na^+ + HCO_3^- \qquad (1\text{-}15)$$

As is shown in Figure 1-4, the only buffer of quantitative importance in the extracellular fluid is the bicarbonate system. It follows, then, that very little of the H^+ that is generated when CO_2 is added to the body — in fact, less than 5% (Fig. 1-7) — can be buffered in the extracellular compartment, i.e., in the plasma and interstitial fluid.

At Onset of Hypoventilation

Blood	Interstitial Fluid	Tissue Cells and Bone
Plasma $CO_2 + H_2O \rightleftharpoons H_2CO_3 \rightleftharpoons H^+ + HCO_3^-$ $H^+ + HCO_3^- + Na^+ + Prot^- \rightleftharpoons$ $Na^+ + HCO_3^- + HProt$ $\boxed{3\%}$	$CO_2 + H_2O \rightleftharpoons H_2CO_3 \rightleftharpoons H^+ + HCO_3^-$ $H^+ + HCO_3^- + 2\,Na^+ + HPO_4^{2-}$ $\rightleftharpoons Na^+ + HCO_3^- + NaH_2PO_4$ $\boxed{\text{Negligible}}$	$HOH \rightleftharpoons H^+$ $\overset{\text{C.A.}}{\rightleftharpoons}$ $CO_2 + OH^- \rightleftharpoons HCO_3^-$ $H^+ + HCO_3^- + K^+ + Prot^- \rightleftharpoons$ $K^+ + HCO_3^- + HProt$ Or: buffered by organic phosphates $\boxed{?11\%}$
Red blood cells $\boxed{29\%}$ $CO_2 \rightarrow CO_2$: Rapid conversion to HCO_3^- and H^+; formation of carbamino hemoglobin; buffering by hemoglobin; exchange of HCO_3^- for Cl^- — all as shown in Figure 1-2.	$Co_2 + H_2O \rightleftharpoons H_2CO_3 \rightleftharpoons HCO_3^- + H^+$ $Na^+ + HCO_3^- \qquad Na^+ + Lac^-$ glucose or glycogen $\xrightarrow{\ \ } HLac \xrightarrow{\ \ } CO_2 + H_2O$ $\boxed{6\%}$	
	$Na^+ + HCO_3^- \rightleftharpoons HCO_3^- + H^+ \rightarrow H^+$ $Na^+ \longrightarrow Na^+ + Prot^- \rightleftharpoons HProt$ Or: buffered by organic phosphates $\boxed{37\%}$	
	$K^+ + HCO_3^- \rightleftharpoons HCO_3^- + H^+ \rightarrow H^+$ $K^+ \longrightarrow K^+ + Prot^- \rightleftharpoons HProt$ Or: buffered by organic phosphates $\boxed{14\%}$	

$CO_2 \rightarrow$

2-5 Days After Onset of Hypoventilation

Renal excretion of H^+ and renal reabsorption of HCO_3^- (see Chap. 2) restore plasma pH to nearly normal value.

Fig. 1-7 : Events that follow the retention of CO_2 (and hence the addition of H^+) due to prolonged alveolar hypoventilation. For the sake of clarity, polyvalent protein anions have been written with a single negative sign. The approximate proportion of the total acid load that is buffered by each mechanism is indicated by the percentages in the rectangles; the figure of 11% is uncertain. Data adapted from Giebisch, G., et al. *J. Clin. Invest.* 34:231, 1955; Pitts, R. F. *Physiology of the Kidney and Body Fluids* (3rd ed.). Chicago: Year Book, 1974; and Masoro, E. J., and Siegel, P. D. *Acid-Base Regulation: Its Physiology, Pathophysiology and the Interpretation of Blood-Gas Analysis* (2nd ed.). Philadelphia: Saunders, 1977.

A second important distinction between the addition of a fixed acid and the addition of the volatile "acid" CO_2 lies in the time required for the acid load to be distributed throughout the major fluid compartments. This process takes hours for a fixed acid (Fig. 1-6), whereas it occurs within minutes for CO_2, which readily diffuses across both the vascular endothelium and the cell membrane.

Hemoglobin, of course, is the most abundant nonbicarbonate buffer in blood (Fig. 1-4). Consequently, a large proportion of the added volatile acid is buffered by the various mechanisms outlined in Figure 1-2. Since CO_2 diffuses so easily across cell membranes, the reactions in erythrocytes occur within seconds after the onset of the disturbance. However, since H^+ is simultaneously being formed in the interstitial compartment (where only small amounts of nonbicarbonate buffers are available), the fall in pH is much greater than would occur if the CO_2 were confined to the blood compartment. The H^+ formed in the extracellular compartment is buffered primarily within cells. Entry of H^+ into cells is accomplished mainly by exchange of H^+ for Na^+ across cell membranes, by H^+ for K^+ exchange in very severe disturbances, and to a lesser extent by H^+ combining with lactate anions within the interstitial fluid. CO_2 rapidly enters all cells (not just erythrocytes), and the H^+ that is formed within cells is also buffered by intracellular nonbicarbonate buffers.

Thus, since the H^+ generated when CO_2 is added cannot be buffered by the bicarbonate system, which exists mainly in the extracellular compartment, all but 3% of the added acid load is buffered either within cells (erythrocytes, tissue cells, and bone) or by lactate, which, as lactic acid, is metabolized within cells. These buffer reactions essentially reach equilibrium within 15 minutes after a change in P_{CO_2} has occurred.

As was the case after the addition of strong acid, so here too the renal response is a relatively slow one, requiring days before a new equilibrium state is reached. In this connection, a third difference between the addition of a fixed acid and the addition of volatile acid should be noted (alluded to earlier, in the last paragraph under Compensatory Responses). The events occurring on the day of adding strong acid (Fig. 1-6) include the rapid second line of defense, in which the excretion of CO_2 "adjusts" the denominator of the Henderson-Hasselbalch equation. Consequently, a near normal plasma pH is attained quickly. However, when the primary acid-base disturbance is respiratory (i.e., when the initial change is in the denominator of the Henderson-Hasselbalch equation; Eq. 1-9), the events occurring at the onset are limited to chemical buffering (Fig. 1-7), and compensatory "adjustment" of the numerator must await the much slower renal process. Hence, primary respiratory disturbances are accompanied by relatively marked deviations of pH during the first few days, and a near normal pH is attained only after a number of days.

Deficit of the Volatile "Acid" CO_2 — Hypocapnia

The responses that set in when alveolar hyperventilation reduces CO_2, and hence H^+ (Eqs. 1-1 and 1-2), are analogous to those depicted in Figure 1-7, except that the reactions proceed in the opposite direction.

Summary

Every day a person produces large quantities of acid. In an adult this amounts to approximately 50 mmoles of nonvolatile acid, derived mainly from dietary proteins and phospholipids, and at least 13,000 mmoles of CO_2 (approximately 90% of which momentarily generates H^+ as the CO_2 is carried in the blood to the lungs, where it is expired [Fig. 1-2]). Despite these large loads of acid the person maintains an alkaline plasma pH, which is crucial to survival.

The concept of CO_2 as an acid is complicated because CO_2 is not a proton donor, nor does H_2CO_3 exist in appreciable amounts in the body fluids. Rather, most of the H^+ that results from the processing of metabolic CO_2 is derived from the dissociation of water (Eq. 1-2 and Fig. 1-2). As metabolic CO_2 enters the blood from tissue cells, this generation of H^+ poses a threat to H^+ balance. The decrease in pH is minimized by a special property of hemoglobin as a buffer, which makes nonoxygenated hemoglobin less acidic than oxygenated hemoglobin. Hence, as arterial capillary blood releases O_2, it can take up a great deal of H^+ without any change in pH (Fig. 1-3). Simultaneously, the CO_2 coming from the cells combines with the OH^- of dissociated water to yield HCO_3^-, and it is mainly in this form that CO_2 is carried to the lungs for excretion.

Acids other than CO_2 do not have a volatile component; they are therefore known as nonvolatile, or "fixed," acids. Normally, such acids are derived from the catabolism primarily of proteins, and to a lesser extent of phospholipids; partial metabolism of carbohydrates and fats as well as loss of HCO_3^- in stool make further contributions. Acid-base balance in the presence of these acids is preserved by three lines of defense: (1) physicochemical buffering, which begins within seconds after introduction of the acid; (2) adjustments in alveolar ventilation, which occur within seconds or minutes; and (3) renal excretion of H^+ and renal reabsorption of HCO_3^-, which may take days to reach completion. The bicarbonate buffer system is especially effective in handling excesses or deficits of fixed acids because (1) it involves an equilibrium with a volatile component, CO_2, which can be regulated by changes in alveolar ventilation, and (2) the substrates of this buffer system, namely, CO_2 and HCO_3^-, are readily available from metabolic processes.

Disturbances of H^+ balance may have either a metabolic origin or a respiratory origin. The first will cause a deviation in the numerator of the Henderson-Hasselbalch equation (Eq. 1-9), which will tend to shift the pH in either an acid or an alkaline direction; the second will lead to a deviation in the denominator, which likewise may shift the pH in either direction. A primary disturbance of one origin (i.e., metabolic or respiratory) is accompanied by a secondary, or compensatory, response of opposite origin (i.e., respiratory or metabolic, respectively), which restores the plasma pH toward normal. Compensations for primary metabolic disturbances are almost instantaneous, whereas those for primary respiratory disturbances become fully effective only after several days. When primary disturbances of metabolic and respiratory origin occur simultaneously in the same individual, the resulting state is known as a mixed disturbance.

A load of acid or alkali calls into play not only the buffers of the plasma but also those of interstitial fluid, cells, and bone. Roughly 50% of fixed acid is buffered within cells. In contrast, the buffering of more than 95% of the H^+ resulting from an excess or deficit of CO_2 (Eqs. 1-1 and 1-2) involves the intracellular space; this is so because this H^+ cannot be buffered by the bicarbonate system (Eqs. 1-14 and 1-15), which is overwhelmingly the predominant buffer of extracellular fluid, i.e., the plasma and interstitium.

The isohydric principle states that all buffer pairs in a common solution are in equilibrium with the same H^+ concentration. The various body fluid compartments do not represent a common solution. Nevertheless, the principle can be applied in most equilibrium states by extending knowledge of the bicarbonate system to assess the approximate status of all buffers in the body.

Problem 1-1

The data below were obtained on each of four patients. Complete the analysis of the acid-base status of each patient by filling in the blank spaces.

Normal arterial values from Table 1-1: pH = 7.38 to 7.42; [HCO$_3^-$] = 22 to 26 mmoles/L; P$_{CO_2}$ = 37 to 43 mm Hg.

Cause of the disturbance	Arterial plasma			Type of disturbance
	pH	P$_{CO_2}$ (mm Hg)	[HCO$_3^-$] (mmoles/L)	
Prolonged vomiting	7.50	49		
Ingestion of NH$_4$Cl[a]		22	10	
Anxiety-hyperventilation syndrome	7.57		21	
Emphysema	7.33	68		

[a]The net effect of ingesting NH$_4$Cl is the addition of hydrochloric acid, by the following reaction:

$$2\ NH_4Cl + CO_2 \rightarrow 2\ H^+ + 2\ Cl^- + H_2O + \underset{\text{urea}}{CO\ (NH_2)_2}$$

Problem 1-2   Outline the sequential steps involved in the development of a steady state of: respiratory acidosis; metabolic alkalosis; and respiratory alkalosis.

What will be the pH of arterial plasma (i.e., acidotic, alkalotic, or unchanged) in the following mixed disturbances: respiratory acidosis plus metabolic acidosis; respiratory acidosis plus metabolic alkalosis; respiratory alkalosis plus metabolic acidosis; and respiratory alkalosis plus metabolic alkalosis?

Appendix: Assembled on pages 27 through 33 are a number of normal
Normal Values   values that are often utilized in dealing with clinical problems of acid-base balance.

Table 1-1 : Normal plasma, serum, or blood concentrations in adults

Substance	Range	Average value usually quoted in treating patients	Comments
Bicarbonate (see also Total CO_2)	23 to 28 mmoles/L 22 to 26 mmoles/L	24 mmoles/L	Venous plasma Arterial blood
Calcium	2.1 to 2.6 mmoles/L	10 mg/dL (2.4 mmoles/L)	Approximately 50% is bound to serum proteins
Carbon dioxide (see Total CO_2)			
Chloride	98 to 106 mmoles/L	100 mmoles/L	
Creatinine	0.5 to 1.5 mg/dL 44 to 133 μmoles/L	1.2 mg/dL (106 μmoles/L)	
Glucose	70 to 100 mg/dL 3.9 to 5.6 mmoles/L	80 mg/dL (4.4 mmoles/L)	Determined in the fasting state; so-called fasting blood sugar (FBS)
Hematocrit (Hct)	40% to 50%	45%	Also frequently called packed cell volume (PCV)
Hydrogen ion [H^+] (arterial)	38 to 43 nmoles/L	40 nmoles/L	Note small units: nanomoles (10^{-9} M)
Lactic acid	0.6 to 1.8 mmoles/L	—	
Lipids: Cholesterol	140 to 310 mg/dL 3.6 to 8.0 mmoles/L	—	
Triglycerides	50 to 150 mg/dL 0.6 to 1.8 mmoles/L	—	
Magnesium	0.6 to 1.1 mmoles/L	0.9 mmole/L	
Osmolality	280 to 295 mosm/kg H_2O	287 mosm/kg H_2O	A value of 300 is often used because it is a round figure that is easy to remember. This approximation does not introduce important quantitative error in most computations for evaluation of fluid and solute balance

Table 1-1 continued on page 28.

Table 1-1 : (Continued)

Substance	Range	Average value usually quoted in treating patients	Comments
Oxygen saturation (arterial)	96% to 100%	—	
Pco₂ (arterial)	37 to 43 mm Hg	40 mm Hg	While breathing room air. Value varies with age
Po₂ (arterial)	75 to 100 mm Hg	—	
pH (arterial)	7.38 to 7.42	7.40	
Phosphorus, inorganic	0.9 to 1.5 mmoles/L	3.5 mg/dL (1.2 mmoles/L)	
Potassium	3.5 to 5.5 mmoles/L	4.5 mmoles/L	
Protein (total)	6 to 8 g/dL	7 g/dL	
Albumin	4 to 5 g/dL	—	
Globulin	2 to 3 g/dL	—	
Sodium	136 to 146 mmoles/L	140 mmoles/L	
Total CO₂ (venous)	22 to 29 mmoles/L	26 mmoles/L	
Urea nitrogen (BUN)	9 to 18 mg/dL 3.2 to 6.4 mmoles/L	12 mg/dL 5 mmoles/L	Measured as the nitrogen contained in urea. Since urea diffuses freely into cells, values for serum, plasma, or whole blood are nearly identical (BUN = blood urea nitrogen). Average value varies with diet; BUN of 18 mg/dL may be normal or may reflect considerable reduction in renal function, depending on intake of protein
Uric acid	3 to 7 mg/dL 0.2 to 0.4 mmoles/L	5 mg/dL (0.3 mmoles/L)	

Table 1-2 : Miscellaneous normal values for adult humans, not given in other figures and tables of this Appendix

	Average value	Comments
Anion gap	12 meq/L	Anion gap = $[Na^+] - ([Cl^-] + [HCO_3^-])$ Normal range: 8 to 16 meq/L
Body surface area	1.73 meter^2	
Body weight	70 kg	
Cerebrospinal fluid (CSF)		For application of Henderson-Hasselbalch equation to CSF
pH	7.30 to 7.38	
$[HCO_3^-]$	22 to 24 mmoles/L	
P_{CO_2}	45 to 50 mm Hg	
pK'	6.13	
Chloride:		
Total body	2,400 mmoles	
Exchangeable	2,000 mmoles	Refers to the amount of Cl^- that is readily miscible with ingested or administered Cl^-
Fluid volumes:		
Total body water (TBW)	60% of body weight	
Intracellular water (ICW)	40% of body weight	Often abbreviated as ICF (intracellular fluid)
Extracellular water (ECW)	20% of body weight	Often abbreviated as ECF (extracellular fluid)
Plasma	4% of body weight	Not calculated separately in most clinical problems but usually considered as the single space of extracellular fluid
Interstitial fluid	16% of body weight	
Glomerular filtration rate (GFR)	125 ml/min	

Table 1-2 continued on page 30.

Table 1-2 : (Continued)

	Average value	Comments
pK′		Selected list of some common physiological compounds
Acetic acid	4.7	
Acetoacetic acid	3.8	
Ammonia	9.2	At 25°C
3-hydroxybutyric acid	4.8	
Creatinine	5.0	
Lactic acid	3.9	
Oxygenated hemoglobin	6.7	
Phosphoric acid	6.8	Refers to HPO_4^{2-} : $H_2PO_4^-$; phosphoric acid has two other pK's
Deoxygenated hemoglobin	7.9	
Potassium:		
Total body	3,500 mmoles	Refers to the total amount of K^+ present in the body
Exchangeable	3,000 mmoles	Refers to the amount of K^+ that is readily miscible with ingested or administered K^+
Sodium:		
Total body	5,000 mmoles	
Exchangeable	3,000 mmoles	Much of the nonexchangeable Na^+ is in bone
Water content of tissues:		
Bone	25%	
Fat	20%	
Muscle	80%	
Plasma	92%	

Table 1-3 : Daily volumes and concentrations of major electrolytes in gastrointestinal secretions, diarrhea, and sweat

	Volume (liters/day)	Electrolyte concentration (mmoles/L)				
		Na$^+$	K$^+$	H$^+$	Cl$^-$	HCO$_3^-$
Saliva: Mean	1.5	30	20	—	31	15
(Range)[a]		(20 to 46)	(16 to 23)		(24 to 44)	(12 to 18)
Gastric juice	2.5	15	10	90	110	0
		(15 to 90)	(4 to 15)	(0 to 100)[c]	(52 to 124)	—
Bile	0.5	140	5	—	105	40
		(120 to 170)	(3 to 12)		(80 to 120)	(30 to 50)
Pancreatic juice	0.7	140	5	—	60	90
		(113 to 153)	(3 to 7)		(54 to 95)	(70 to 110)
Small intestine	1.5	120	5	—	110	35
		(72 to 158)	(4 to 7)		(70 to 127)	(20 to 40)
Diarrhea	1.0 to 1.5[b]	130	10	—	95	20
		(120 to 140)	(5 to 15)		(90 to 100)	(15 to 30)
Sweat	0 to 3.0[b]	50	5	—	50	0
		(18 to 97)	(1 to 15)		(18 to 97)	—

[a]In many clinical situations (e.g., vomiting, diarrhea, or surgical drainage of intestinal secretions) volume and electrolyte concentrations are not determined. In lieu of precise measurements, the physician then proceeds to treat the patient on the basis of estimates. The mean values listed serve as a useful guide for such estimates; normal ranges are also given in order to emphasize that there is wide variability.

[b]May be greater.

[c]Hydrogen ion concentration varies inversely with sodium concentration, depending on the stimulus for acid secretion.

Data compiled from M. G. Rosenfeld (ed.), *Manual of Medical Therapeutics* (22nd ed.). Boston: Little, Brown, 1977; J. L. Gamble, *Chemical Anatomy, Physiology, and Pathology of Extracellular Fluid* (6th ed.). Cambridge, Mass.: Harvard University Press, 1954; A. I. Arieff, in M. H. Maxwell and C. R. Kleeman (eds.), *Clinical Disorders of Fluid and Electrolyte Metabolism* (2nd ed.). New York: McGraw-Hill, 1972.

32

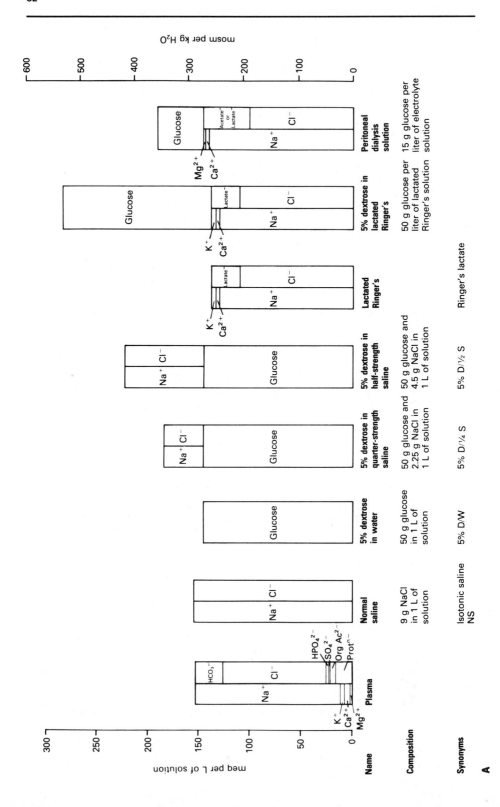

Solutions Used Occasionally or Rarely

Solution	Dextrose g per L of solution	Na$^+$	K$^+$	Ca^{2+}	Cl$^-$	Alcohol ml per L of solution	Approximate Osmolality mosm/kg H$_2$O
		mmoles per L of solution					
10% dextrose	100						548
20% dextrose	200						1096
5% dextrose in 5% alcohol	50					50	1361
5% dextrose in 0.9% NaCl	50	154			154		565
10% dextrose in 0.9% NaCl	100	154			154		839
Ringer's solution		147	4	2.5	156		295
0.45% NaCl		77			77		146
5% NaCl		856			856		1617

B

Concentrated Solutions Used as Additives

Name	Volume of commercial ampule ml	Concentration per ml	Comments
Ammonium chloride	30	3 mmoles NH$_4^+$ [a]	Always diluted before use; used as an acidifying agent, net effect being addition of HCl
Calcium chloride	10	0.7 mmoles Ca^{2+}	
Calcium gluconate	10	0.25 mmoles Ca^{2+}	
Glucose, 50%	50	0.5 g dextrose	
Magnesium sulfate	2	2.0 mmoles Mg^{2+}	
Mannitol	50	0.25 g mannitol	
Potassium chloride	10	2 mmoles K$^+$	
Potassium phosphate	30	2 mmoles K$^+$	
Sodium acetate	30	3 mmoles acetate	Used as an alkalinizing agent, acetate being metabolized to bicarbonate
Sodium bicarbonate	50	1 mmole HCO$_3^-$	
Sodium chloride	40	2.5 mmoles Na$^+$	Always diluted before use
Sodium lactate	20	2.5 mmoles lactate	Used as an alkalinizing agent, lactate being metabolized to bicarbonate

[a]Ion listed is the one for which solution is given to patient.

C

Fig. 1-8 : Composition of various parenteral solutions. The ones that are most commonly used in many hospitals are illustrated by the bar graphs (A). For comparison, the composition of normal plasma is given in the bar graph on the left. Other solutions that are also available commercially, but used only occasionally or rarely, have been listed in the box on the left (B). The commercial forms of various concentrated solutions are given in the box on the right (C); note that their concentrations are expressed *per milliliter*, which emphasizes the fact that most of these solutions must be diluted prior to use.

Selected
References

General

Cohen, J. J., and Kassirer, J. P. *Acid-Base.* Boston: Little, Brown, 1982.

Davenport, H. W. *The ABC of Acid-Base Chemistry* (6th ed.). Chicago: University of Chicago Press, 1974.

Davis, R. P. Logland: A Gibbsian view of acid-base balance. *Am. J. Med.* 42:159, 1967.

Dejours, P. (ed.). Symposium on interaction of intra- and extracellular acid-base balance. *Respir. Physiol.* 33:1, 1978.

Fencl, V. Distribution of H^+ and HCO_3^- in Cerebral Fluids. In B. H. Siesjö and S. C. Sørensen (eds.), *Ion Homeostasis of the Brain. The Regulation of Hydrogen and Potassium Ion Concentrations in Cerebral Intra- and Extracellular Fluids.* New York: Academic, 1971.

Henderson, L. J. *Blood: A Study in General Physiology.* New Haven: Yale University Press, 1928.

Hills, A. G. *Acid-Base Balance: Chemistry, Physiology, Pathophysiology.* Baltimore: Williams & Wilkins, 1973.

Huckabee, W. E. Henderson vs. Hasselbalch. *Clin. Res.* 9:116, 1961.

Masoro, E. J., and Siegel, P. D. *Acid-Base Regulation: Its Physiology, Pathophysiology and the Interpretation of Blood-Gas Analysis* (2nd ed.). Philadelphia: Saunders, 1977.

Nahas, G. G. (ed.). Current concepts of acid-base measurement. *Ann. N.Y. Acad. Sci.* 133:1, 1966.
Two of the papers in this symposium, Terminology of Acid-Base Disorders and Statement on Acid-Base Terminology, have also been reproduced in Ann. Intern. Med. *63: 873 and 885, 1965.*

Nattie, E. E. Ionic mechanisms of cerebrospinal fluid acid-base regulation. *J. Appl. Physiol: Respiratory, Environmental and Exercise Physiol.* 54:3, 1983.

Nattie, E. E., and Romer, L. CSF HCO_3^- regulation in isosmotic conditions. The role of brain Pco_2 and plasma HCO_3^-. *Respir. Physiol.* 33:177, 1978.

Pappenheimer, J. R. The ionic composition of cerebral extracellular fluid and its relation to the control of breathing. *Harvey Lect.* 61:71, 1965.

Pitts, R. F. *Physiology of the Kidney and Body Fluids* (3rd ed.). Chicago: Year Book, 1974.

Rector, F. C., Jr. (ed.). Symposium on acid-base homeostasis. *Kidney Int.* 1:273, 1972.

Robin, E. D., Bromberg, P. A., and Cross, C. E. Some aspects of the evolution of vertebrate acid-base regulation. *Yale J. Biol. Med.* 41:448, 1969.

Roos, A., and Boron, W. F. Intracellular pH. *Physiol. Rev.* 61:296, 1981.

Rose, B. D. *Clinical Physiology of Acid-Base and Electrolyte Disorders* (2nd ed.). New York: McGraw-Hill, 1984.

Schwartz, W. B., and Cohen, J. J. The nature of the renal response to chronic disorders of acid-base equilibrium. *Am. J. Med.* 64: 417, 1978.

Schwartz, W. B., and Relman, A. S. A critique of the parameters used in evaluation of acid-base disorders. "Whole-blood buffer base" and "standard bicarbonate" compared with blood pH and plasma bicarbonate concentration. *N. Engl. J. Med.* 268:1382, 1963.

Siesjö, B. H. The regulation of cerebrospinal fluid pH. *Kidney Int.* 1:360, 1972.

Siggard-Andersen, O. *The Acid-Base Status of the Blood* (4th ed.). Baltimore: Williams & Wilkins, 1974.
This book provides an excellent description of the basic physical chemistry of blood buffers. It also describes in detail the concept of "base excess" in evaluating acid-base disorders. Although we do not recommend Professor Siggard-Andersen's system, the reader is directed to this source for the description of this approach.

Valtin, H. *Renal Function: Mechanisms Preserving Fluid and Solute Balance in Health* (2nd ed.). Boston: Little, Brown, 1983. Chapters 9 and 10.

Van Ypersele de Strihou, C., and Frans, A. The respiratory response to chronic metabolic alkalosis and acidosis in disease. *Clin. Sci. Mol. Med.* 45:439, 1973.

Wiggins, P. M. Intracellular pH and the structure of cell water. *J. Theor. Biol.* 37:363, 1972.

Winters, R. W. (ed.). *The Body Fluids in Pediatrics.* Boston: Little, Brown, 1973.

Winters, R. W., Engel, K., and Dell, R. B. *Acid-Base Physiology in Medicine: A Self-Instruction Program* (3rd ed.). Boston: Little, Brown, 1982.

Buffering

Adrogué, H. J., Brensilver, J., Cohen, J. J., and Madias, N. E. Influence of steady-state alterations in acid-base equilibrium on the fate of administered bicarbonate in the dog. *J. Clin. Invest.* 71:867, 1983.

Bergstrom, W. H., and Wallace, W. M. Bone as a sodium and potassium reservoir. *J. Clin. Invest.* 33:867, 1954.

Bidani, A., Crandall, E., and Forster, R. Analysis of postcapillary pH changes in blood in vivo after gas exchange. *J. Appl. Physiol: Respiratory, Environmental and Exercise Physiol.* 44:770, 1978.

Boron, W. F. Intracellular pH transients in giant barnacle muscle fibers. *Am. J. Physiol.* 233: (Cell Physiol. 2):C61, 1977.

German, B., and Wyman, J., Jr. The titration curves of oxygenated and reduced hemoglobin. *J. Biol. Chem.* 117:533, 1937.

Giebisch, G., Berger, L., and Pitts, R. F. The extrarenal response to acute acid-base disturbances of respiratory origin. *J. Clin. Invest.* 34:231, 1955.

Kilmartin, J. V., and Rossi-Bernardi, L. Interaction of hemoglobin with hydrogen ions, carbon dioxide, and organic phosphates. *Physiol. Rev.* 58:836, 1973.

Lemann, J., Jr., and Lennon, E. J. Role of diet, gastrointestinal tract and bone in acid-base homeostasis. *Kidney Int.* 1:275, 1972.

Lemann, J., Jr., Lennon, E. J., Goodman, A. D., Litzow, J. R., and Relman, A. S. The net balance of acid in subjects given large loads of acid or alkali. *J. Clin. Invest.* 44:507, 1965.

Pitts, R. F. Mechanisms for stabilizing the alkaline reserves of the body. *Harvey Lect.* 48:172, 1953.

Sendroy, J., Jr., Seelig, S., and Van Slyke, D. D. Studies of acidosis. XXII. Application of the Henderson-Hasselbalch equation to human urine. *J. Biol. Chem.* 106:463, 1934.

Swan, R. C., Axelrod, D. R., Seip, M., and Pitts, R. F. Distribution of sodium bicarbonate infused into nephrectomized dogs. *J. Clin. Invest.* 34:1795, 1955.

Swan, R. C., and Pitts, R. F. Neutralization of infused acid by nephrectomized dogs. *J. Clin. Invest.* 34:205, 1955.

Van Slyke, D. D. On the measurement of buffer values and on the relationship of buffer values to the dissociation constant of the buffer and the concentration and reaction of the buffer solution. *J. Biol. Chem.* 52:525, 1922.

Yoshimura, H., Fujimoto, M., Okumura, O., Sugimoto, J., and Kuwada, T. Three-step regulation of acid-base balance of body fluid after acid load. *Jpn. J. Physiol.* 11:109, 1961.

2 : Role of Kidneys in Acid-Base Balance: Renal Excretion of H^+ and Conservation of HCO_3^-

Role of Kidneys in H^+ Balance

It is clear from Chapter 1 that the maintenance of a normal plasma pH depends on the preservation of a normal *ratio* between the weak acid and conjugate base components of each of the body buffers. According to the isohydric principle (Eq. 1-13) these ratios can be determined precisely for all plasma buffers, from knowledge of the bicarbonate buffer system in plasma. Except for the slight correction necessitated by the Gibbs-Donnan effect (see The Isohydric Principle, Chap. 1), the plasma bicarbonate system will also reflect the ratio of all interstitial buffers; and, as stated previously, in the steady state of most acid-base disturbances, any change in the plasma bicarbonate system will be accompanied by qualitatively similar changes of the intracellular buffers. It thus follows that regulating the ratio of the concentration of HCO_3^- to that of Pco_2 in plasma (Eq. 1-9) will tend to regulate the ratio of all other buffer pairs.

The weak acid component of the plasma bicarbonate buffer system is regulated as Pco_2 through alveolar ventilation (Eq. 1-9). Preservation of the conjugate base, HCO_3^-, is accomplished by the kidneys. This task involves two processes: (1) the reabsorption of virtually all the HCO_3^- that was filtered, and (2) the reclamation of the HCO_3^- that was consumed in buffering fixed acids (Eq. 1-12). The latter process is accomplished through the excretion of an equivalent amount of H^+ into the urine. The renal replenishment of HCO_3^- stores through excretion of H^+ was mentioned in Chapter 1, under Third Line of Defense.

Reabsorption of Filtered HCO_3^-

Like Na^+ and other small solutes, HCO_3^- is freely filtered by the glomeruli. Because of the very high glomerular filtration rate (GFR), the filtered load of HCO_3^- — i.e., the amount of HCO_3^- that is filtered into the tubules (GFR $\times$ $P_{HCO_3}-$) — is very great; in an adult human it amounts to approximately 4,500 mmoles per day. If even a small portion of this quantity were to be excreted in the urine, the normal stores of this important buffer would quickly be exhausted. This eventuality is prevented by avid tubular reabsorption of HCO_3^-, which normally exceeds 99.9% of the filtered load; that is, normally only about 2 mmoles of HCO_3^- are excreted in the urine.

Mechanism for
Reabsorption of
Filtered HCO₃⁻

In a series of classic studies conducted in the 1940s, R. F. Pitts and his colleagues showed conclusively that much of the acid that is excreted gets into the urine not by glomerular filtration but by tubular secretion. They reasoned that the source of this acid must be largely or exclusively carbon dioxide (Eq. 1-1), and they strengthened their thesis by demonstrating that inhibition of the enzyme carbonic anhydrase (C.A.; Eq. 1-2) greatly reduces or abolishes the amount of acid that can be secreted. They suggested, furthermore, that acid is secreted in the form of H^+ ion in exchange for Na^+, rather than as molecular acid. A tremendous amount of subsequent experimental work by numerous investigators has proved those suggestions to be largely correct. Figure 2-1 embodies them; although the schema varies in some details from that first drawn up by Pitts, the essence of his proposal has stood the test of time. Even now, some of the details are not fully settled.

Within tubular cells, water is split into H^+ and OH^- (Fig. 2-1). (As stated in conjunction with Fig. 1-2, this reaction, rather than Eq.

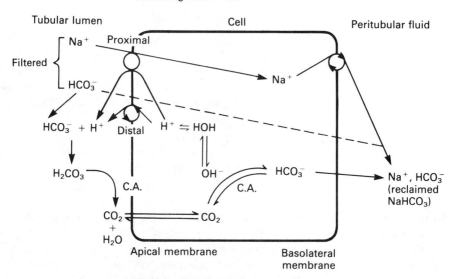

Proximal tubule 80–90%
Loop of Henle ≈ 5%
Distal tubule ≈ 3%
Collecting duct 1–2%

Fig. 2-1 : Mechanism for the reabsorption of filtered HCO₃⁻ in rats. C.A. stands for carbonic anhydrase. In the proximal tubule, but not in the distal tubule or the collecting duct, tubular fluid is exposed to this enzyme, which is located in the luminal membrane. Under normal conditions, essentially all HCO₃⁻ is reabsorbed by the reaction of filtered HCO₃⁻ with secreted H⁺, rather than by the direct route indicated by the interrupted arrow. In the proximal tubule H⁺ secretion is linked to Na⁺ entry into the cell, whereas in the distal tubule and collecting duct H⁺ secretion is relatively independent of Na⁺ entry. The percentages (derived from measurements in rats) indicate the proportion of filtered HCO₃⁻ that is reabsorbed in each part of the nephron.

1-1, is chosen for the generation of H^+ because the hydroxylation of CO_2 [Eq. 1-2] is thought to be the reaction that is catalyzed by carbonic anhydrase.) The H^+ is secreted into the tubular lumen by two mechanisms: (1) through a process that is linked to the passive entry of Na^+ into the cell, and (2) through an active transport pump. The first mechanism predominates in the proximal tubule, the second in later parts of the nephron. In the tubular lumen the secreted H^+ combines with filtered HCO_3^- to form H_2CO_3, which is converted to CO_2 and water. In the proximal tubule this conversion occurs within milliseconds under the influence of carbonic anhydrase. The arrow for this step is deflected toward the cell to indicate that the carbonic anhydrase resides in the luminal cell membrane; tubular fluid is thus exposed to the enzyme even though carbonic anhydrase is not found in tubular fluid as such. The CO_2 that is formed within the lumen diffuses into the cell, where it combines with the OH^- that results from the dissociation of water, to form HCO_3^- — again, under the influence of carbonic anhydrase. This HCO_3^- then diffuses into the peritubular fluid and blood, along with Na^+, which is actively transported across the peritubular (also called basolateral) membrane. In addition, some filtered HCO_3^- can be reabsorbed directly; normally, however, this process contributes little to overall bicarbonate reabsorption.

Note the net effects of the processes illustrated in Figure 2-1. For every H^+ that is secreted, a filtered Na^+ is reabsorbed (even when not by direct linkage), and with the Na^+ a HCO_3^- is returned to peritubular fluid and blood. The mechanism shown in Figure 2-1 thus accomplishes the important task of *reclaiming virtually all the filtered HCO_3^-*. Note that it is not a mechanism for excreting H^+; to the extent that the CO_2 formed within the tubular lumen from secreted H^+ returns to the cell, ultimately to form more H^+ through hydroxylation, no net secretion of H^+ takes place.

The proportion of the total filtered HCO_3^- that disappears from tubular fluid in each of the major parts of the nephron is shown at the top of Figure 2-1. These values were derived in experimental animals by means of clearance ratios, i.e., from knowledge of the concentrations of HCO_3^- and inulin in arterial plasma and tubular fluid. The results show that, by far, the filtered HCO_3^- is reabsorbed largely in the proximal tubule, mainly in its early part.

Factors Influencing the Rate at Which Filtered HCO_3^- Is Reabsorbed

The rate at which filtered HCO_3^- is returned to the blood can be affected by a number of factors, which often interact. Among the more important influences are the following: (1) the amount of HCO_3^- presented to the tubules; (2) the size of the extracellular fluid volume; (3) the arterial P_{CO_2}; and (4) certain hormones.

1. Figure 2-2A shows a so-called titration curve for HCO_3^-, obtained by producing stepwise increments in plasma HCO_3^- concentration. As the plasma HCO_3^- concentration is varied, so is the filtered load of this ion (GFR $\times$ $P_{HCO_3^-}$) and hence the amount of HCO_3^- that is presented to the tubules for reabsorption. Provided that other variables, such as the volume of extracellular fluid (see below), are held constant, the amount of HCO_3^- that is reabsorbed is almost the same as the load that is filtered into the tubules. The mechanism for this effect has not been clarified, save that the rate of HCO_3^- reabsorption appears to be closely linked to that of Na^+ reabsorption, especially in the early proximal tubule. It may thus be in part a consequence of the need to conserve Na^+ and maintain the extracellular fluid volume.

2. Also shown in Figure 2-2A is the effect of the extracellular fluid volume. When that volume is expanded markedly — as by infusing either $NaHCO_3$ or NaCl — the reabsorption of filtered HCO_3^- is decreased, and the converse holds when the extracellular fluid volume is contracted. Although there appears to be a plateau for HCO_3^- reabsorption during exaggerated volume expansion, the true maximal reabsorptive rate for HCO_3^- (Tm) is much higher. Again, the mechanisms by which extracellular fluid volume influences HCO_3^- reabsorption have not been fully clarified. Given the close linkage between Na^+ and HCO_3^- reabsorption in the proximal tubule, the effect probably involves reduction in proximal Na^+ reabsorption, which is known to occur during volume expansion and for which some of the mechanisms are known.

3. The influence of arterial P_{CO_2} is shown in Figure 2-2B. As P_{CO_2} is lowered (as by hyperventilation), the reabsorption of filtered HCO_3^- is decreased, and as P_{CO_2} is raised (as by alveolar hypoventilation), HCO_3^- reabsorption is increased. The effect is more marked during chronic than during acute alterations of the P_{CO_2}. The mechanisms that drive this response have not been conclusively identified. Some think that they involve the supply of more CO_2 substrate to the cell (Fig. 2-1), but others consider this view too simplistic and have suggested that CO_2 may stimulate the secretory pump for H^+ or some other, related transport step. Whatever the mechanisms, it is clear from Equation 1-9 that the response will tend to correct the blood pH during a primary respiratory disturbance.

4. An increased plasma concentration of adrenal corticosteroids, as in Cushing's syndrome, leads to increased reabsorption of filtered HCO_3^-, and the converse occurs during adrenal cortical insufficiency (Addison's disease). This effect is not mediated solely by influencing the renal handling of Na^+ or

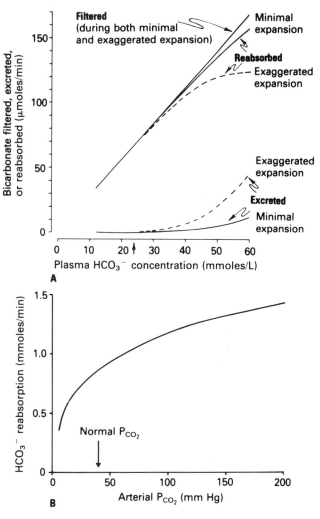

Fig. 2-2 : A. Titration curves for HCO_3^- in normal rats. Virtually all the filtered HCO_3^- is reabsorbed unless the extracellular fluid volume is markedly expanded, in which case the reabsorptive rate is reduced despite a similar increase in the filtered load. The arrow indicates a normal arterial plasma HCO_3^- concentration (see Table 1-1). Adapted from Purkerson, M. L., et al., *J. Clin. Invest.* **48**:1754, 1969.

B. Changes in the rate of HCO_3^- reabsorption in dogs as the arterial Pco_2 is either lowered through hyperventilation or raised through breathing gas mixtures containing increased concentrations of CO_2. Adapted from Rector, F. C., Jr., et al., *J. Clin. Invest.* **39**:1706, 1960.

of K^+; it also involves an enhancement of H^+ secretion by adrenal steroids.

Acute infusion of parathyroid hormone decreases the reabsorption of HCO_3^-, but sustained administration of the hormone may increase HCO_3^- reabsorption slightly.

Replenishment of Depleted HCO_3^- Stores

It was pointed out at the beginning of Chapter 1 that in persons whose diet is fairly high in protein there is a net daily production of nonvolatile (fixed) acids. These include sulfuric acid, resulting from protein catabolism; phosphoric acid, which is produced chiefly during the catabolism of phospholipids; and organic acids. These acids are buffered by the following types of reactions:

$$2H^+ + SO_4^{2-} + 2Na^+ + 2HCO_3^- \rightleftharpoons 2Na^+ + SO_4^{2-} + 2H_2O + 2CO_2 \nearrow \qquad (2\text{-}1)$$

$$2H^+ + HPO_4^{2-} + 2Na^+ + 2HCO_3^- \rightleftharpoons 2Na^+ + HPO_4^{2-} + 2H_2O + 2CO_2 \nearrow \qquad (2\text{-}2)$$

The CO_2 is eliminated via the lungs, as indicated by the diagonal arrows, and the two neutral salts, Na_2SO_4 and Na_2HPO_4, are filtered into Bowman's space. If these neutral salts were excreted in the urine, the body would soon become depleted of $NaHCO_3$, the main extracellular buffer that is utilized in neutralizing the fixed acids. The kidneys prevent such depletion of $NaHCO_3$, by two means: (1) the excretion of NH_4^+, and (2) the excretion of titratable acid (T.A.). In both operations HCO_3^-, newly formed within renal tubular cells, is absorbed into the peritubular blood along with Na^+ that was filtered.

Excretion of Titratable Acid (T.A.)

As tubular fluid is acidified — by the reabsorption of HCO_3^- and consequent decline of its concentration in tubular fluid (Eq. 1-9) — secreted H^+ ions combine with other filtered buffers in the tubular fluid. As part of the latter process, the neutral salt Na_2HPO_4 is converted to the acid salt NaH_2PO_4 (for a depiction of this process, see Fig. 2-5). The amount of strong base required to titrate the acid urine back to a pH of 7.40 (which is the approximate pH of the glomerular filtrate) is *equal to the amount of titratable acid that was excreted in the urine.* (Other filtered buffers, such as creatinine and the organic anions, citrate, acetate, and 3-hydroxy butyrate, are also titrated, but they normally contribute only a trivial amount to T.A. because of their low concentration and low pK'.)

The probable schema for the formation of urinary T.A. is shown in Figure 2-3. The major reaction that generates the secreted H^+ is thought to be the dissociation of water; the OH^- that is simul-

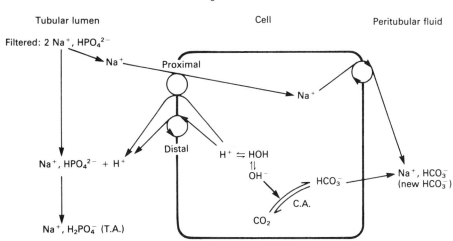

Fig. 2-3 : **Mechanism whereby titratable acid (T.A.) is created and newly formed HCO_3^- is added to the blood along with a reabsorbed Na^+. In the proximal tubule the CO_2 comes largely from the tubular lumen (see Fig. 2-1); in more distal parts of the nephron it may come mainly from cellular metabolism. C.A. = carbonic anhydrase.**

taneously liberated combines with intracellular CO_2, under the catalysis of carbonic anhydrase, to form the new HCO_3^- that is added to peritubular fluid and blood. Within the tubular lumen the secreted H^+ combines with filtered 2 Na^+, HPO_4^{2-} to form Na^+, $H_2PO_4^-$, which is excreted as T.A. in the urine. The second filtered Na^+ that is liberated in this reaction is reabsorbed to combine with HCO_3^- that was newly formed within the cell. These reactions occur in all major parts of the nephron; in fact, they take place in the same cells as the schema depicted in Figure 2-1. But note that, whereas the net effect in Figure 2-1 is to recapture virtually all the filtered HCO_3^-, the net effect in Figure 2-3 is to *replenish the blood with one HCO_3^- for every HCO_3^- that was consumed in buffering fixed H^+* (Eq. 2-2).

Factors Affecting the Rate of T.A. Excretion

Two factors influence the rate at which T.A. is excreted: the availability of urinary buffers and the pK' of those buffers.

1. The effect of availability of urinary buffers is illustrated in Figure 2-4A, in which an increased supply of urinary buffer is reflected on the abscissa as increased excretion of phosphate. As more buffer is made available, more T.A. is excreted.

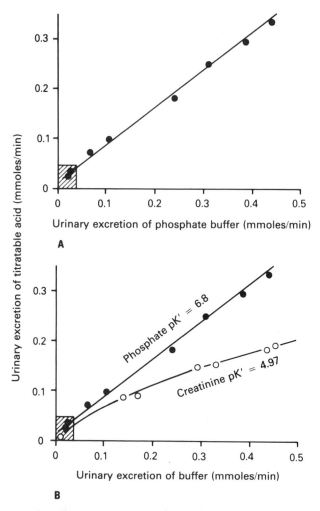

Fig. 2-4 : Two experiments in humans, illustrating factors that influence the rate at which titratable acid (T.A.) is excreted in the urine. The experiments were done on the same person, who was either in a state of normal H^+ balance or in metabolic acidosis induced by ingesting NH_4Cl. This salt leads to acidosis through net addition of hydrochloric acid, as in the following reaction:

$$2\,NH_4Cl + CO_2 \rightleftharpoons 2\,HCl + H_2O + \underset{\text{urea}}{CO(NH_2)_2}$$

A. In this experiment the subject was in mild metabolic acidosis (plasma pH = 7.37; plasma HCO_3^- concentration = 14 mmoles/liter; urinary pH = 4.5). The points enclosed in the shaded rectangle represent excretion of endogenous phosphate; all other points were obtained during intravenous infusion of inorganic phosphate at pH 7.40.

B. Differences in the amount of H^+ excreted as T.A. when phosphate is the main urinary buffer or when creatinine is the main buffer. The points for phosphate are those illustrated in (A); those for creatinine (open circles) are data obtained in the same subject (plasma pH = 7.38; plasma HCO_3^- concentration = 13 mmoles/liter; urinary pH = 5.1) when creatinine was infused intravenously instead of phosphate.

Both graphs are slightly modified from Schiess, W. A., Ayer, J. L., Lotspeich, W. D., and Pitts, R. F., *J. Clin. Invest.* 27:57, 1948. It is of interest that the subject for these experiments was Dr. Pitts, whose work has contributed so much to our understanding of the renal regulation of acid-base balance.

The explanation involves a limiting concentration gradient for the transport of H^+ by renal cells. Besides HCO_3^-, inorganic phosphate is the main urinary buffer. At a pH of 7.4, at which it is filtered into Bowman's space, phosphate exists mainly as $2\,Na^+$, HPO_4^{2-} (Fig. 2-5). As the urine becomes more acid, the phosphate is converted to Na^+, $H_2PO_4^-$, which is the main urinary T.A. The minimal urinary pH is about 4.4, probably because collecting duct cells cannot transport H^+ against a concentration gradient exceeding about 1:1,000. When this minimal pH is attained (as it was in the experiment shown in Fig. 2-4A), virtually all the urinary phosphate is in the Na^+, $H_2PO_4^-$ form, and addition of even minute amounts of H^+ would then lead to a precipitous drop in pH (Fig. 2-5). Hence, under these conditions more H^+ can be excreted as T.A. only if more phosphate is filtered, i.e., only if the availability of more phosphate buffer in the tubular fluid permits the acceptance of more H^+ without a further drop in pH. This requirement was met by raising the plasma phosphate concentration and is reflected in the steadily increasing phosphate excretion in Figure 2-4A.

2. A buffer is most effective within $\pm$ 1.0 pH unit of its pK (Fig. 1-5). Hence, given the normal pH of glomerular filtrate of about 7.4, phosphate with a pK of 6.8 can initially accept much more H^+ per unit drop in tubular fluid pH than can another buffer with a lower pK (Fig. 2-5). Furthermore, if the pK of the other buffer is close to the minimal urinary pH of 4.4, the total amount of H^+ that a quantum of that buffer can accept over the normal range of tubular fluid pH will be less than the amount accepted by the same quantum of phosphate. This fact is reflected in Figure 2-4B. Per millimole of buffer in the urine, more H^+ can be excreted as T.A. when phosphate is the main urinary buffer than when creatinine, with a pK of 4.97, is the main urinary buffer.

This example illustrates the difference between *urinary acidification* and *H^+ excretion.* The ability to reduce the pH of urine, which is acidification, does not necessarily tell us much about the amount of H^+ being excreted. Note that the ordinate in Figure 2-5 shows the amount of H^+ excreted per quantum of HPO_4^{2-} presented for titration. Therefore, if, say, 10 times more HPO_4^{2-} were to traverse the tubular system (as was done in the experiment shown in Fig. 2-4A, by increasing the filtered load of phosphate), the amount of H^+ excreted could rise tenfold without any change in urinary pH.

Excretion of Ammonium — Nonionic Diffusion

If the formation of T.A. were the only mechanism for excreting H^+, the amount of H^+ that could be eliminated in the urine would be severely limited by the amount of phosphate and, to a

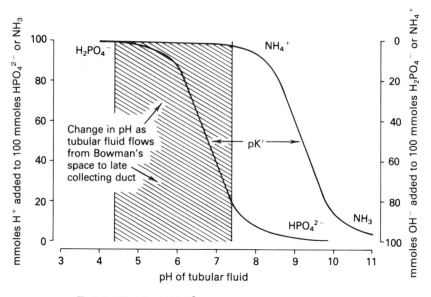

Fig. 2-5 : Titration of HPO_4^{2-} and NH_3 by H^+ as the pH of tubular fluid is decreased from 7.4 in Bowman's space to 4.4 in the late collecting duct. With a pK of 9.2, the NH_3/NH_4^+ system is a relatively poor buffer in the pH range that ordinarily exists in tubular fluid (shaded area). Nevertheless, a great deal of H^+ can be excreted as NH_4^+ because the supply of NH_3 from tubular cells is potentially plentiful (see Table 2-1). Adapted from Valtin, H., *Renal Dysfunction: Mechanisms Involved in Fluid and Solute Imbalance.* Boston: Little, Brown, 1979.

lesser extent, of other buffers that are filtered. As soon as the titration curve for phosphate would shift to the formation of H_3PO_4 from Na^+, $H_2PO_4^-$, urinary pH would fall below 4.4 (Fig. 2-5) and no more H^+ could be secreted by the renal tubular cells (see point 1 above); yet it can be shown that much more H^+ than that which appears as T.A. can be excreted in the urine, even though the urine pH does not fall below the minimum value of 4.4. It is therefore apparent that an additional mechanism exists for the excretion of H^+.

The observation that in acidosis there is a rise not only in urinary T.A. but also in urinary NH_4^+ raised the suspicion that NH_3 might be the additional acceptor for H^+, as by the following reaction:

$$H^+ + Cl^- + NH_3 \rightleftharpoons NH_4^+ + Cl^- \qquad (2\text{-}3)$$

Note that the H^+ is incorporated into the neutral salt NH_4Cl, so that this reaction satisfies the requirement of excreting H^+ without a further decrease in urinary pH; in other words, neutral ammonium salts are not titratable acids.

The suspicion that the ammonia/ammonium system was involved was strengthened by the following findings: that the con-

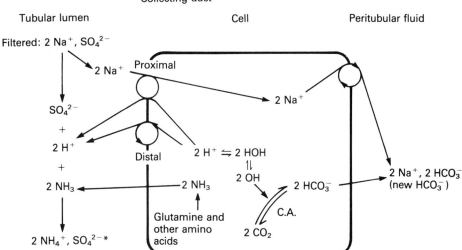

Fig. 2-6 : Mechanism for the renal excretion of NH_4^+. Glutamine and other amino acid substrates come from the blood and enter the cell from the peritubular and luminal side. In the proximal tubule the CO_2 comes largely from the tubular lumen (see Fig. 2-1); in more distal parts of the nephron it may come mainly from cellular metabolism. C.A. = carbonic anhydrase.

*The excretion of 2 NH_4^+, SO_4^{2-} is shown here, rather than that of NH_4^+, Cl^-, to indicate how the filtered Na_2SO_4 (which is derived from buffering of H_2SO_4; Eq. 2-1) is handled. In the kidney, however, there is no special linkage between sulfate and ammonium, and NH_4^+ is, in fact, excreted mainly as the chloride salt.

centration of NH_3 is higher in renal venous blood than in renal arterial blood; that the delivery of NH_3 to the kidney from arterial blood is too low to account for the amount of NH_4^+ that is excreted in the urine; and that the concentration of NH_3 in arterial blood is unchanged in severe acidosis when urinary NH_4^+ excretion is greatly enhanced. The conclusion seemed inescapable that NH_3 must be produced within the kidney, and that it is excreted in the urine as NH_4^+ after accepting a H^+.

The probable mechanism for the urinary excretion of NH_4^+ is illustrated in Figure 2-6. Here, too, the hydroxylation of CO_2 under the influence of carbonic anhydrase is thought to be the source of the HCO_3^- that is added to peritubular fluid and blood; and the H^+ that is simultaneously evolved from the splitting of water is secreted into the tubular lumen.

NH_3 is derived from glutamine and other amino acids in the blood, which are broken down within the renal tubular cells. This nonionized form is lipid-soluble and thus is freely diffusible across virtually the entire cell membrane, which is composed largely of fat. NH_3 therefore diffuses passively down its concentration gradient into the tubular lumen. The pK' of the NH_3/NH_4^+ buffer system is about 9.2 (Table 1-2). Therefore, at the pH of tubular fluid, H^+ avidly combines with NH_3, so that the system exists almost entirely in the NH_4^+ form (see Fig. 2-5). This ionized form, unlike NH_3, is not lipid-soluble and therefore traverses the cell membrane much less readily, since its transit is confined to the aqueous channels. Consequently, the NH_4^+ is trapped within the tubular lumen, and it is then excreted in the form of neutral salts, such as NH_4Cl or $(NH_4)_2SO_4$. (For reasons explained in the legend to Figure 2-6, the excretion of 2 NH_4^+, SO_4^{2-}, rather than of NH_4^+, Cl^-, is shown; it should be realized, however, that NH_4^+ is excreted mainly as the chloride salt [Fig. 2-8], since Cl^- is far and away the most abundant anion in normal urine.) In the process, the HCO_3^- that was newly formed within the tubular cell is added to the blood along with the filtered Na^+ that is reabsorbed. As was the case with the excretion of T.A., the net result of the NH_4^+ mechanism is thus *the excretion of H^+, the replenishment of the body HCO_3^- stores, and the reabsorption of filtered Na^+*. These reactions probably occur in all parts of the nephron.

Nonionic Diffusion

The process by which the lipid-soluble, nonionized moiety of a buffer pair (e.g., NH_3) can readily diffuse across a cell membrane, while the lipid-insoluble, ionized member (e.g., NH_4^+) cannot, is known as nonionic diffusion or diffusion-trapping. Note that in the case of the urinary excretion of NH_4^+ this process aids the secretion of NH_3 and hence the excretion of H^+. The moment that NH_3 enters the tubular lumen, virtually all of it is converted to NH_4^+ (Figs. 2-5 and 2-6), so that a constant "sink" for the continued diffusion of NH_3 is maintained. This positive feedback, so to speak, is the more effective, the lower the urinary pH (Fig. 2-7); i.e., the secretion of NH_3 and, hence, the excretion of H^+ are most efficient in acidosis, when more H^+ needs to be eliminated.

Nonionic diffusion is a common biological phenomenon that has important clinical applications in promoting the urinary excretion of weak acids and weak bases. An example of the utilization of this principle in the treatment of phenobarbital poisoning is given in Problem 2-1 and the corresponding Answer.

Control of Renal NH_3 Production and Excretion

At least four factors influence the amount of NH_3 that is produced and secreted into the tubular lumen: the pH of the urine, the chronicity of acidosis, the relative rates of flow of peritubular blood and tubular fluid, and the total body stores of K^+.

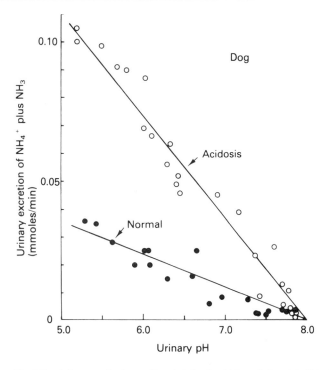

Fig. 2-7 : Influence of urinary pH and state of acid-base balance on the urinary excretion of total ammonia (NH₄⁺ plus NH₃). All the data were obtained from one dog. At the beginning of each experiment the urinary pH was about 5.2, both in normal H⁺ balance and after 48 hours of metabolic acidosis. The urinary pH was then gradually increased in each instance by infusing NaHCO₃ intravenously. From Pitts, R. F., *Fed. Proc.* 7:418, 1948.

1. The influence of urinary pH was discussed above and is depicted in Figure 2-7. During both normal H⁺ balance and states of metabolic acidosis, there is an inverse relationship between the urinary pH and the amount of total ammonia (i.e., NH_3 plus NH_4^+) that is excreted in the urine. The more acid the urine, the greater the proportion of total ammonia that exists in the ionized form (see Fig. 2-5); hence, the more acid the urine, the lower the urinary concentration of NH_3 and the greater the concentration difference promoting the passive diffusion of NH_3 from renal tubular cell into the tubular lumen. Since this NH_3 is immediately converted to NH_4^+ within the tubular lumen, it cannot diffuse back into the cell, but is instead excreted.

2. The influence of the duration of acidosis is also illustrated in Figure 2-7, which shows that, *at any given urinary pH,* the rate of total ammonia excretion is higher in acidosis (especially when acidosis is present for hours or days) than during normal H⁺ balance. It is important to note that this does not

involve the explanation given in paragraph 1, since the difference can be detected at the same urinary pH. Rather, the explanation involves an increased renal production of NH_3, which is an important adaptive mechanism that permits increased excretion of H^+ during acidosis. Despite a great deal of work, it is not yet known precisely how this adaptation comes about.

3. When tubular fluid has the same pH as blood, which it usually does not (Fig. 2-5), the flow rates of these two solutions determine the rates of NH_3 diffusion into them. Since peritubular blood flow is much greater than the flow of tubular fluid, blood would carry off the NH_3 more rapidly and thereby maintain a more favorable sink for the further diffusion of NH_3. Normally, however, when tubular fluid is acid, about 75% of the NH_3 produced within the renal cells diffuses into the tubular lumen, and about 25% into the blood. Under the admittedly unusual conditions when the pH of the two fluids is equal or when the pH is higher in tubular fluid than in blood, most of the renal NH_3 diffuses into the blood; when the urinary pH is 8, virtually no NH_3 enters the tubular fluid (Fig. 2-7).

4. Potassium depletion stimulates NH_3 production, and K^+ loading suppresses production of NH_3. The mechanism for this effect is not known. The effect does not, however, alter total acid excretion; in K^+-depleted subjects, for example, the excretion of T.A. falls as that of NH_4^+ rises.

Relative Excretion Rates of Titratable Acid and Ammonium

Table 2-1 shows rates of H^+ excretion as NH_4^+ and as T.A., both in the normal situation and in two disease states that are characterized by disturbances of H^+ balance. It was pointed out at the beginning of Chapter 1 that in a normal person whose diet is relatively high in protein there is a net daily production of 40 to

Table 2-1 : Relative excretion rates of ammonium and titratable acid (T.A.) in healthy persons and in two disease states that are accompanied by primary metabolic acidosis

Condition	mmoles of urinary H^+ per day
Health	
H^+ combined with NH_3	30 to 50
H^+ as T.A.	10 to 30
Diabetic acidosis	
H^+ combined with NH_3	300 to 500
H^+ as T.A.	75 to 250
Chronic renal disease	
H^+ combined with NH_3	0.5 to 15
H^+ as T.A.	2 to 20

From Pitts, R. F. Science 102:49;81, 1945. Published with permission.

60 mmoles of nonvolatile acids. In the steady (equilibrium) state, this amount of acid is excreted as H^+ combined with NH_3 or as T.A.; more H^+ is normally excreted as NH_4^+ than as T.A.

During uncontrolled diabetes mellitus there is an overproduction of nonvolatile acids, mainly 3-OH butyric acid. This leads to primary metabolic acidosis, which, as discussed earlier, increases the urinary excretion of NH_4^+ (Fig. 2-7). The pK′ of 3-OH butyrate is slightly less than that of creatinine (Table 1-2); hence 3-OH butyrate is ordinarily a less effective urinary buffer than are phosphate and creatinine (Figs. 2-4B and 2-5). During diabetic acidosis, however, the endogenous production and, hence, the filtered load of 3-OH butyrate are so great that in this condition 3-OH butyrate becomes the main urinary buffer, and T.A. appears primarily as 3-OH butyric acid. Nevertheless, the data in Table 2-1 during diabetic acidosis show that the potential supply of NH_3 as a urinary buffer is enormous, and considerably greater than that of buffers that form titratable acids.

During chronic renal disease, which is accompanied by a marked decrease in the amount of functioning renal tissue, there may be a reduction in both forms of fixed-acid excretion, depending largely on the protein content of the patient's diet. The reduction, however, is relatively much greater for NH_3 than for T.A. This is so because the rate of T.A. excretion depends largely on the urinary excretion of buffer (Fig. 2-4A), which may remain normal until renal disease is far advanced; formation of NH_4^+, however, depends on the renal cellular production of NH_3, which is greatly curtailed as the amount of functioning renal tissue is reduced.

Summary

The kidneys play a major role in the regulation of H^+ balance by maintaining the normal body store (and hence concentration) of HCO_3^-, and by excreting the H^+ that is derived from the daily production of nonvolatile (fixed) acids. Preservation of HCO_3^- stores is accomplished through: (1) the reabsorption of virtually all the HCO_3^- that is filtered (Fig. 2-1); and (2) the formation of new HCO_3^- within renal cells and addition of this HCO_3^- to the blood (Figs. 2-3 and 2-6). Net excretion of H^+ also occurs via two mechanisms: (1) the excretion of titratable acid (T.A.; Fig. 2-3), which is quantified as the amount of strong base that must be added to acid urine in order to return the pH of that urine to that of blood; and (2) the excretion of neutral NH_4^+ salts after H^+ has combined with secreted NH_3 (Fig. 2-6). The last process involves the principle of nonionic diffusion, or diffusion-trapping.

In this chapter each of these processes was described separately, both for the sake of clarity and to account quantitatively for the various aspects of H^+ balance. It must be realized, how-

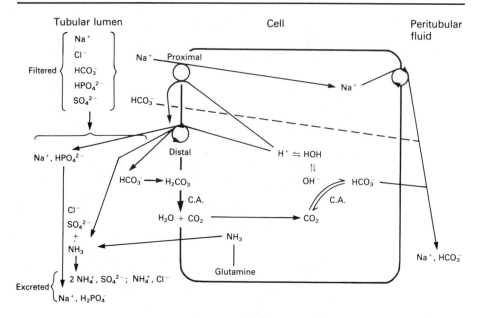

Fig. 2-8 : Summary diagram, showing the various means by which the kidneys help maintain H⁺ balance. Although each mechanism occurs in virtually all parts of the nephron, some processes predominate in one part (e.g., titration of filtered HCO₃⁻ in the proximal tubule) while others predominate in another part (e.g., titration of NH₃ in the collecting duct). The diagram is a simplification, which does not show some unresolved questions. It is likely, for example, that the hydroxylation of CO₂ is not the only source for secreted H⁺, and that carbonic anhydrase has a function in the peritubular membrane (possibly to enhance HCO₃⁻ transport) as well as in the luminal membrane and cytoplasm.

ever, that the major mechanisms take place simultaneously in the same cells, that the secreted H^+ comes from a common cellular pool, and that the kidney does not distinguish H^+ that is destined to combine with filtered HCO_3^- from that which will combine with HPO_4^{2-} or NH_3; nor will it distinguish a reabsorbed HCO_3^- that was generated within the cell from one that was initially filtered, or a reabsorbed Na^+ that came from $NaHCO_3$ from one that came from the chloride, phosphate, sulfate, or some other filtered salt. In order to emphasize these points, the processes are summarized in Figure 2-8, where they are shown as occurring in a single tubular cell.

The rate at which filtered HCO_3^- is reabsorbed is affected by the filtered load of HCO_3^-, expansion or contraction of the extracellular fluid, the arterial P_{CO_2}, and certain hormones, notably adrenal corticosteroids and parathyroid hormone. The rate of T.A. excretion is influenced principally by the availability of buffers within the tubular fluid and the pK' of those buffers. The rate of urinary NH_4^+ excretion is governed by the pH of the urine, the chronicity of acidosis, the relative flow rates of peritubular blood as compared to tubular fluid (especially when the urinary pH is equal to or higher than the pH of plasma), and the body stores of K^+.

Normally, about three-fourths of the daily endogenous load of nonvolatile acid is excreted as NH_4^+, and the remainder as T.A. The potential supply of NH_3 as a H^+ acceptor is very large, rising as much as tenfold in states such as diabetic acidosis. Since NH_3 is generated within renal cells, however, this adaptive mechanism may be greatly curtailed in chronic renal disease, when the total amount of functioning renal tissue is reduced.

Problem 2-1

Utilization of the Principle of Nonionic Diffusion in the Treatment of Phenobarbital Poisoning. A 23-year-old woman is admitted to the emergency ward in coma and with a history of having ingested a large amount of phenobarbital. Her respirations are shallow, and her systemic blood pressure is somewhat low. The patient is several times incontinent of urine, which probably means that urine production is adequate despite the mild hypotension.

After instituting measures to re-establish normal respiration and to support the systemic circulation, the attending physician begins efforts to hasten excretion of the phenobarbital. Although much of the drug is metabolized in the liver, as much as 30% of the total dose may be excreted unchanged by the kidney. The compound, having a molecular weight of 232 daltons, enters the tubular system mainly through filtration, and it is then passively reabsorbed. When the urine flow is adequate, as in this patient, forced diuresis and alkalinization of the urine constitute the preferred mode of therapy.

The attending physician therefore begins to infuse mannitol and $NaHCO_3$ intravenously. Mannitol, with a molecular weight of 182 daltons, is freely filtered but is reabsorbed very poorly or not at all. Hence, by contributing to the osmolality of tubular fluid, mannitol inhibits the passive reabsorption of water and initiates a diuresis. The consequent dilution of phenobarbital in the tubular fluid decreases its passive reabsorption. The infusion of $NaHCO_3$ will alkalinize the urine, shift the titration of the weak acid phenobarbital (pK' of 7.2) toward the ionized form (Fig. 2-5), and thereby diminish the passive reabsorption of the phenobarbital through the process of nonionic diffusion.

It is determined that the concentration of total phenobarbital (i.e., the nonionized plus the ionized form) in this patient is 10 mg per 100 ml of plasma. About 40% of phenobarbital is bound to plasma proteins. On admission, the plasma pH is 7.3 (primary respiratory acidosis due to depression of the respiratory center) and urinary pH is 5.2. After infusion of $NaHCO_3$ and correction of the alveolar hypoventilation through assisted ventilation, the plasma pH is 7.7 and the urinary pH is 8.2. The pK' of the phenobarbital system is 7.2. Given these facts, complete the table below.

	Total unbound phenobarbital in plasma (mg/100 ml)	Ratio of unbound phenobarbital: [Ionized] / [Nonionized]	Plasma concentration of unbound phenobarbital		
			Ionized	Nonionized
			(mg/100 ml)	
Plasma, pH 7.3				
Plasma, pH 7.7				
Urine, pH 5.2	—		—	—
Urine, pH 8.2	—		—	—

Problem 2-2

Is NH_4Cl a titratable acid (T.A.)? Defend your answer, utilizing Figure 2-5.

Problem 2-3

Theoretically, to what extent could urine be acidified solely by reabsorbing filtered HCO_3^-, without secretion of H^+? *Hint:* Apply the Henderson-Hasselbalch equation (Eq. 1-9), using a pK' of 6.1 and a P_{CO_2} for urine of 40 mm Hg.

Selected References

General

Alpern, R. J., Warnock, D. G., and Rector, F. C., Jr. Renal Acidification Mechanisms. In B. M. Brenner and F. C. Rector, Jr. (eds.), *The Kidney* (3rd ed.). Philadelphia: Saunders, 1985.

Baruch, S. B. (ed.). Symposium on renal metabolism. *Med. Clin. North Am.* 59:505, 1975.
This symposium was held in honor of Dr. Robert F. Pitts, whose work elucidated many of the mechanisms by which the kidney regulates H^+ balance. In the beautifully written and learned Dedication that introduces this volume, Dr. Erich E. Windhager describes the qualities that made Dr. Pitts a greatly admired teacher and investigator.

Cohen, J. J., and Kassirer, J. P. *Acid-Base.* Boston: Little, Brown, 1982. Chapter 4.

Gennari, F. J., and Cohen, J. J. Role of the kidney in potassium homeostasis: Lessons from acid-base disturbances. *Kidney Int.* 8:1, 1975.

Malnic, G., and Giebisch, G. Cellular Aspects of Renal Tubular Acidification. In G. Giebisch (ed.), *Transport Organs,* vol IVA. New York: Springer-Verlag, 1979.

Malnic, G., and Steinmetz, P. R. Transport processes in urinary acidification. *Kidney Int.* 9:172, 1976.

Masoro, E. J., and Siegel, P. D. *Acid-Base Regulation: Its Physiology, Pathophysiology and the Interpretation of Blood Gas Analysis* (2nd ed.). Philadelphia: Saunders, 1977.

Pitts, R. F. Mechanisms for stabilizing the alkaline reserves of the body. *Harvey Lect.* 48:172, 1953.

Rector, F. C., Jr. Sodium, bicarbonate, and chloride absorption by the proximal tubule. *Am. J. Physiol.* 244 (Renal Fluid Electrolyte Physiol. 13):F461, 1983.

Rector, F. C., Jr. (ed.). Symposium on acid-base homeostasis. *Kidney Int.* 1:273, 1972.

Steinmetz, P. R. Cellular mechanisms of urinary acidification. *Am. J. Physiol.* (Renal Fluid Electrolyte Physiol.), 1986, in press.

Tannen, R. L. Control of acid excretion by the kidney. *Annu. Rev. Med.* 31:35, 1980.

H^+ Secretion

Al-Awqati, Q. H^+ transport in urinary epithelia. *Am. J. Physiol.* 235 (Renal Fluid Electrolyte Physiol. 4):F77, 1978.

Aronson, P. S. Mechanisms of active H^+ secretion in the proximal tubule. *Am. J. Physiol.* 245 (Renal Fluid Electrolyte Physiol. 14):F647, 1983.

Brodsky, W. A. (ed.). Anion and proton transport. *Ann. N.Y. Acad. Sci.* 341:1, 1980.

Cohen, L. H., and Steinmetz, P. R. Control of active proton transport in turtle urinary bladder by cell pH. *J. Gen. Physiol.* 76:381, 1980.

DeSousa, R. C., Harrington, J. T., Ricanati, E. S., Shelkrot, J. W., and Schwartz, W. B. Renal regulation of acid-base equilibrium during chronic administration of mineral acid. *J. Clin. Invest.* 53:465, 1974.

DuBose, T. D., Jr. Application of the disequilibrium pH method to investigate the mechanisms of urinary acidification. *Am. J. Physiol.* 245 (Renal Fluid Electrolyte Physiol. 14):F535, 1983.

Forte, J. G., Warnock, D. G., and Rector, F. C., Jr. (eds.). *Hydrogen Ion Transport in Epithelia.* New York: Wiley, 1984.
This volume is the proceedings of a symposium that dealt with basic research into H^+ transport. The last section of the symposium deals with the kidney; the earlier sections review fundamental aspects of H^+ secretion, including gastrointestinal H^+ secretion.

Gennari, F. J., Caflisch, C. R., Johns, C., Maddox, D. A., and Cohen, J. J. P_{CO_2} measurements in surface proximal tubules and peritubular capillaries of the rat kidney. *Am. J. Physiol.* 242 (Renal Fluid Electrolyte Physiol. 11):F78, 1982.

Gennari, F. J., Johns, C., and Caflisch, C. R. Effect of benzolamide on pH in the proximal tubules and peritubular capillaries of the rat kidney. *Pflügers Arch.* 387:69, 1980.

Gottschalk, C. W., Lassiter, W. E., and Mylle, M. Localization of urine acidification in the mammalian kidney. *Am. J. Physiol.* 198:581, 1960.

Green, H. H., Steinmetz, P. R., and Frazier, H. S. Evidence for proton transport by turtle bladder in presence of ambient bicarbonate. *Am. J. Physiol.* 218:845, 1970.

Halperin, M. L., and Jungas, R. L. Metabolic production and renal disposal of hydrogen ions. *Kidney Int.* 24:709, 1983.

Harris, R. C., Seifter, J. L., and Brenner, B. M. Adaptation of Na^+-H^+ exchangers in renal microvillus membrane vesicles. *J. Clin. Invest.* 74:1979, 1984.

Hulter, H. N., Licht, J. H., Bonner, E. L., Jr., Glynn, R. D., and Sebastian, A. Effects of glucocorticoid steroids on renal and systemic acid-base metabolism. *Am. J. Physiol.* 239 (Renal Fluid Electrolyte Physiol. 8):F30, 1980.

Koeppen, B. M., and Steinmetz, P. R. Basic mechanisms of urinary acidification. *Med. Clin. North Am.* 67:753, 1983.

Malnic, G., de Mello-Aires, M., and Giebisch, G. Micropuncture study of renal tubular hydrogen ion transport in the rat. *Am. J. Physiol.* 222:147, 1972.

Murer, H., Hopfer, U., and Kinne, R. Sodium/proton antiport in brush-border–membrane vesicles isolated from rat small intestine and kidney. *Biochem. J.* 154:597, 1976.

Rector, F. C., Jr., Carter, N. W., and Seldin, D. W. The mechanism of bicarbonate reabsorption in the proximal and distal tubules of the kidney. *J. Clin. Invest.* 44:278, 1965.

Schwartz, G. J., and Al-Awqati, Q. Carbon dioxide causes exocytosis of vesicles containing H^+ pumps in isolated perfused proximal and collecting tubules. *J. Clin. Invest.* 75:1638, 1985.

Schwartz, J. H., Finn, J. T., Vaughan, G., and Steinmetz, P. R. Distribution of metabolic CO_2 and the transported ion species in acidification by turtle bladder. *Am. J. Physiol.* 226:283, 1974.

Tan, S.-C., Goldstein, M. B., Stinebaugh, B. J., Chen, C.-B., Gougoux, A., and Halperin, M. L. Studies on the regulation of hydrogen ion secretion in the collecting duct in vivo: Evaluation of factors that influence the urine minus blood P_{CO_2} difference. *Kidney Int.* 20:636, 1981.

Terao, N., and Tannen, R. L. Characterization of acidification by the isolated perfused rat kidney: Evidence for adaptation by the distal nephron to a high bicarbonate diet. *Kidney Int.* 20:36, 1981.

Ullrich, K. J., Rumrich, G., and Baumann, K. Renal proximal tubular buffer-(glycodiazine) transport. Inhomogeneity of local transport rate, dependence on sodium, effect of inhibitors and chronic adaptation. *Pflügers Arch.* 357:149, 1975.

Vieira, F. L., and Malnic, G. Hydrogen ion secretion by rat renal cortical tubules as studied by an antimony microelectrode. *Am. J. Physiol.* 214:710, 1968.

Walser, M., and Mudge, G. H. Renal Excretory Mechanisms. In C. L. Comar and F. Bronner (eds.), *Mineral Metabolism.* New York: Academic, 1960.

Warnock, D. G., and Rector, F. C., Jr. Proton secretion by the kidney. *Annu. Rev. Physiol.* 41:197, 1979.

Ziegler, T. W., Fanestil, D. D., and Ludens, J. H. Influence of transepithelial potential difference on acidification in the toad urinary bladder. *Kidney Int.* 10:279, 1976.

Conservation of
HCO_3^-

Alpern, R. J., Cogan, M. G., and Rector, F. C., Jr. Effects of extracellular fluid volume and plasma bicarbonate concentration on proximal acidification in the rat. *J. Clin. Invest.* 71:736, 1983.

Alpern, R. J., Cogan, M. G., and Rector, F. C., Jr. Flow dependence of proximal tubular bicarbonate absorption. *Am. J. Physiol.* 245 (Renal Fluid Electrolyte Physiol. 14):F478, 1983.

Bernstein, H., Atherton, L. J., and Deen, W. M. Axial heterogeneity and filtered-load dependence of proximal bicarbonate reabsorption. *Biophys. J.,* 1986, in press.

Boylan, J. W., Antkowiak, D. E., and Calkins, J. Maximum rates of bicarbonate reabsorption by the dogfish kidney. *Bull. Mount Desert Island Biol. Lab.* 13:17, 1973.

Burg, M. B., and Green, N. Bicarbonate transport by isolated perfused rabbit proximal convoluted tubules. *Am. J. Physiol.* 233 (Renal Fluid Electrolyte Physiol. 2):F307, 1977.

Cogan, M. G., and Alpern, R. J. Regulation of proximal bicarbonate reabsorption. *Am. J. Physiol.* 247 (Renal Fluid Electrolyte Physiol. 16):F387, 1984.

Cogan, M. G., Maddox, D. A., Lucci, M. S., and Rector, F. C., Jr. Control of proximal bicarbonate reabsorption in normal and acidotic rats. *J. Clin. Invest.* 64:1168, 1979.

Hellman, D. E., Au, W. Y. W., and Bartter, F. C. Evidence for a direct effect of parathyroid hormone on urinary acidification. *Am. J. Physiol.* 209:643, 1965.

Hulter, H. N. Effects and interrelationships of PTH, Ca^{2+}, Vitamin D, and P_i in acid-base homeostasis. *Am. J. Physiol.* 248 (Renal Fluid Electrolyte Physiol. 17):F739, 1985.

Kurtz, I., Maher, T., Hulter, H. N., Schambelan, M., and Sebastian, A. Effect of diet on plasma acid-base composition in normal humans. *Kidney Int.* 24:670, 1983.

Maddox, D. A., Atherton, L. J., Deen, W. M., and Gennari, F. J. Proximal HCO_3^- reabsorption and the determinants of tubular and capillary P_{CO_2} in the rat. *Am. J. Physiol.* 247 (Renal Fluid Electrolyte Physiol. 16):F73, 1984.

Maddox, D. A., and Gennari, F. J. Load dependence of HCO_3^- and H_2O reabsorption in the early proximal tubule of the Munich-Wistar rat. *Am. J. Physiol.* 248 (Renal Fluid Electrolyte Physiol. 17):F113, 1985.

Madias, N. E., Wolf, C. J., and Cohen, J. J. Regulation of acid-base equilibrium in chronic hypercapnia. *Kidney Int.* 27:538, 1985.

Maren, T. H. Chemistry of the renal reabsorption of bicarbonate. *Can. J. Physiol. Pharmacol.* 52:1041, 1974.

McKinney, T. D., and Burg, M. B. Bicarbonate and fluid absorption by renal proximal straight tubules. *Kidney Int.* 12:1, 1977.

McKinney, T. D., and Burg, M. B. Bicarbonate transport by rabbit cortical collecting tubules. Effect of acid and alkali loads in vivo on transport in vitro. *J. Clin. Invest.* 60:766, 1977.

Mello-Aires, M., and Malnic, G. Peritubular pH and P_{CO_2} in renal tubular acidification. *Am. J. Physiol.* 228:1766, 1975.

Pitts, R. F., Ayer, J. L., and Schiess, W. A. The renal regulation of acid-base balance in man. III. The reabsorption and excretion of bicarbonate. *J. Clin. Invest.* 28:35, 1949.

Purkerson, M. L., Lubowitz, H., White, R. W., and Bricker, N. S. On the influence of extracellular fluid volume expansion on bicarbonate reabsorption in the rat. *J. Clin. Invest.* 48:1754, 1969.

Rector, F. C., Jr., Seldin, D. W., Roberts, A. D., Jr., and Smith, J. S. The role of plasma CO_2 tension and carbonic anhydrase activity in the renal reabsorption of bicarbonate. *J. Clin. Invest.* 39:1706, 1960.

Titratable Acid and
Ammonia

Balagura-Baruch, S. Renal Metabolism and Transfer of Ammonia. In C. Rouiller and A. F. Muller (eds.), *The Kidney,* vol. III. New York: Academic, 1971.

Buerkert, J., Martin, D., and Trigg, D. Segmental analysis of the renal tubule in buffer production and net acid formation. *Am. J. Physiol.* 244 (Renal Fluid Electrolyte Physiol. 13):F442, 1983.

Burch, H. B., Chan, A. W. K., Alvey, T. R., and Lowry, O. H. Localization of glutamine accumulation and tubular reabsorption in rat nephron. *Kidney Int.* 14:406, 1978.

Cheema-Dhadli, S., and Halperin, M. L. Role of the mitochondrial anion transporters in the regulation of ammoniagenesis in renal cortex mitochondria of the rabbit and rat. *Eur. J. Biochem.* 99:483, 1979.

Glabman, S., Klose, R. M., and Giebisch, G. Micropuncture study of ammonia excretion in the rat. *Am. J. Physiol.* 205:127, 1963.

Goldstein, L. Adaptation of renal ammonia production to metabolic acidosis: A study in metabolic regulation. *Physiologist* 23 (1):19, 1980.

Goldstein, L. Ammonia production and excretion in the mammalian kidney. *Int. Rev. Physiol* 11:283, 1976.

Good, D. W., and Burg, M. B. Ammonia production by individual segments of the rat nephron. *J. Clin. Invest.* 73:602, 1984.

Good, D. W., and Knepper, M. A. Ammonia transport in the mammalian kidney. *Am. J. Physiol.* 248 (Renal Fluid Electrolyte Physiol. 17):F459, 1985.

Kamm, D. E., and Strope, G. L. The effects of acidosis and alkalosis on the metabolism of glutamine and glutamate in renal cortex slices. *J. Clin. Invest.* 51:1251, 1972.

Malnic, G., Mello-Aires, M., DeMello, G. B., and Giebisch, G. Acidification of phosphate buffer in cortical tubules of rat kidney. *Pflügers Arch.* 331:275, 1972.

Nash, T. P., Jr., and Benedict, S. R. The ammonia content of the blood, and its bearing on the mechanism of acid neutralization in the animal organism. *J. Biol. Chem.* 48:463, 1921.

Pitts, R. F. Control of renal production of ammonia. *Kidney Int.* 1:297, 1972.

Pitts, R. F. Production and Excretion of Ammonia in Relation to Acid-Base Regulation. In J. Orloff and R. W. Berliner (eds.), *Handbook of Physiology,* Section 8, Renal Physiology. Washington, D.C.: American Physiological Society, 1973.

Pitts, R. F. The renal regulation of acid base balance with special reference to the mechanism for acidifying the urine. *Science* 102:81, 1945.

Pitts, R. F., and Alexander, R. S. The nature of the renal tubular mechanism for acidifying the urine. *Am. J. Physiol.* 144:239, 1945.

Pitts, R. F., and Lotspeich, W. D. Factors governing the rate of excretion of titratable acid in the dog. *Am. J. Physiol.* 147:481, 1946.

Relman, A. S., and Narins, R. G. The control of ammonia production in the rat. *Med. Clin. North Am.* 59:583, 1975.

Ross, B., and Guder, W. G. (eds.). *Biochemical Aspects of Renal Function.* Oxford: Pergamon, 1980. Chap. 2.

Simon, E., Martin, D., and Buerkert. J. Contributions of individual superficial nephron segments to ammonium handling in chronic metabolic acidosis in the rat. Evidence for ammonia disequilibrium in the renal cortex. *J. Clin. Invest.* 76:855, 1985.

Tannen, R. L. Ammonia metabolism. *Am. J. Physiol.* 235 (Renal Fluid Electrolyte Physiol. 4):F265, 1978.

Tannen, R. L., and Ross, B. D. Ammoniagenesis by the isolated perfused rat kidney: The critical role of urinary acidification. *Clin. Sci. Mol. Med.* 56:353, 1979.

Ullrich, K. J., and Papavassiliou, F. Bicarbonate reabsorption in the papillary collecting duct of rats. *Pflügers Arch.* 389:271, 1981.

Carbonic Anhydrase

Cogan, M. G., Maddox, D. A., Warnock, D. G., Lin, E. T., and Rector, F. C., Jr. Effect of acetazolamide on bicarbonate reabsorption in the proximal tubule of the rat. *Am. J. Physiol.* 237 (Renal Fluid Electrolyte Physiol. 6):F447, 1979.

Dobyan, D. C., and Bulger, R. E. Renal carbonic anhydrase. *Am. J. Physiol.* 243 (Renal Fluid Electrolyte Physiol. 12):F311, 1982.

Karlmark, B., Ågerup, B., and Wistrand, P. J. Renal proximal tubular acidification. Role of brush-border and cytoplasmic carbonic anhydrase. *Acta Physiol. Scand.* 106:145, 1979.

Lönnerholm, G., and Wistrand, P. J. Carbonic anhydrase in the human kidney: A histochemical and immunocytochemical study. *Kidney Int.* 25:886, 1984.

Lucci, M. S., Tinker, J. P., Weiner, I. M., and DuBose, T. D., Jr. Function of proximal tubule carbonic anhydrase defined by selective inhibition. *Am. J. Physiol.* 245 (Renal Fluid Electrolyte Physiol. 14):F443, 1983.

Maren, T. H. Carbon dioxide equilibria in the kidney: The problems of elevated carbon dioxide tension, delayed dehydration, and disequilibrium pH. *Kidney Int.* 14:395, 1978.

Maren, T. H. Carbonic anhydrase: Chemistry, physiology, and inhibition. *Physiol. Rev.* 47:595, 1967.

Norby, L. H., Bethencourt, D., and Schwartz, J. H. Dual effect of carbonic anhydrase inhibitors on H^+ transport by the turtle bladder. *Am. J. Physiol.* 240 (Renal Fluid Electrolyte Physiol. 9):F400, 1981.

Tinker, J. P., Coulson, R., and Weiner, I. M. Dextran-bound inhibitors of carbonic anhydrase. *J. Pharmacol. Exp. Ther.* 218:600, 1981.

Wistrand, P. J., and Kinne, R. Carbonic anhydrase activity of isolated brush border and basal-lateral membranes of renal tubular cells. *Pflügers Arch.* 370:121, 1977.

Nonionic Diffusion   Hill, J. B. Salicylate intoxication. *N. Engl. J. Med.* 288:1110, 1973.

Levy, G., Lampman, T., Kamath, B. L., and Garrettson, L. K. Decreased serum salicylate concentrations in children with rheumatic fever treated with antacid. *N. Engl. J. Med.* 293:323, 1975.

Milne, M. D., Scribner, B. H., and Crawford, M. A. Non-ionic diffusion and the excretion of weak acids and bases. *Am. J. Med.* 24:709, 1958.

Waddell, W. J., and Butler, T. C. The distribution and excretion of phenobarbital. *J. Clin. Invest.* 36:1217, 1957.

3 : Approach to the Patient: Useful Tools

In the preceding chapters we set forth the basic principles that underlie the maintenance of acid-base balance in health. The remainder of this book is devoted to illustrating how these principles are applied to the analysis of acid-base *im*balances in disease and how such analysis then leads to logical management of patients. In this chapter we describe the general approach to clinical problems; in subsequent chapters we discuss specific clinical examples, drawn largely from our own experience.

Analyzing Clinical Problems

Basic Questions

Three questions need to be answered for every patient:

1. What is the primary disturbance, or, in the case of a mixed disturbance, what are the primary disturbances (see p. 10, Concept of Metabolic and Respiratory Disturbances)?
2. Is the compensatory response (see p. 11, Compensatory Responses) quantitatively appropriate?
3. What is the cause of each primary disturbance? (The causes are tabulated in each of the following chapters, where they will be discussed in detail.)

In many patients, especially those with chronic illnesses, the presence of a specific acid-base disorder will be suggested by the clinical presentation. In the majority of patients with acute illnesses, however, a disorder of acid-base equilibrium is brought to the attention of a physician by an abnormality in the venous total CO_2, which is routinely measured as part of the "serum electrolytes." In all instances the physician should use the tools discussed below, in conjunction with findings from the history and physical examination, to arrive at the answers to the questions listed above. (*Note:* Some laboratory tests are performed on serum and others on plasma. Since the distinction is negligible for most routine laboratory purposes, we shall use the terms "serum" and "plasma" interchangeably.)

Diagrams and Other Aids

Over the years numerous aids have been proposed, all designed to help the physician understand problems of acid-base balance. Such aids include various diagrams of the Henderson-Hasselbalch equation (Eq. 1-9, p. 5) or nomograms of the buffer characteristics of blood, which yield a designation of the patient's acid-base disturbance; sometimes terms such as "base excess" and "standard bicarbonate" are introduced. Several points about these aids should be made: (1) All are merely graphical depictions or logical extensions of the Henderson-Hasselbalch (Eq. 1-9) or the Henderson (Eq. 3-4) relationships. None, therefore, adds a new concept. In fact, the use of these aids *without full understanding* of the above relationships may obscure the dynamics of acid-base disorders. (2) Some diagrams assume, wrongly, that the buffer response of the intact organism to added CO_2 is the same as that observed in blood alone. (3) Use of the aids, especially in inexperienced hands, tends to underplay the importance of the history and physical examination of the patient. (4) An "automatic" application of the aids, without concurrent and independent analysis, permits handling problems in H^+ balance without necessarily understanding the mechanisms involved. Such usage may therefore lead to the wrong diagnosis and, hence, to incorrect treatment. (5) Many of the aids pass through periods of popularity and then fall out of vogue; furthermore, different institutions use different aids. For all these reasons, we shall not invoke these aids in this book.

Laboratory Tests

In current clinical practice, problems of acid-base balance are evaluated in one of two ways (listed in order of increasing complexity) or through a combination of the two:

1. *Venous Total CO_2.* This measurement is obtained routinely as part of the serum electrolytes (i.e., Na^+, K^+, Cl^-, total CO_2). The total CO_2 includes not only HCO_3^- but also dissolved CO_2 and H_2CO_3. Since the last two combined usually constitute less than 10% of the total, the venous total CO_2 value is only 1 to 3 mmoles per liter higher than the concurrent arterial HCO_3^- concentration ($[HCO_3^-]$). The venous total CO_2 is thus a good approximation of the arterial $[HCO_3^-]$, which satisfies the Henderson-Hasselbalch equation (Eq. 1-9).

   Knowledge of the electrolytes also permits calculation of the anion gap, which is discussed as the last of the four useful tools.

2. *Arterial pH and P_{CO_2} (Blood Gas Analysis).* An arterial blood sample is obtained anaerobically, and pH and P_{CO_2} are measured on this sample. The corresponding $[HCO_3^-]$ is then usually computed and reported automatically by the laboratory by using the Henderson-Hasselbalch equation (Eq. 1-9).

Ordinarily, the $[H^+]$ for the same sample (in nmoles/L) is computed and reported as well. This test, then, supplies the three variables of the Henderson-Hasselbalch or the Henderson equation (Eq. 3-4) on a single arterial sample. In addition, Po_2 is measured on the same sample. Knowledge of this variable — as well as of the hematocrit (Hct), hemoglobin (Hgb), and oxygen saturation, which are often supplied — adds useful information about the respiratory system, which plays such an important role in acid-base balance and imbalance.

Some physicians use an intermediate step, in which the pH and Pco_2 are measured on venous blood. This test can supply ballpark values, mainly to tell whether the patient has an acidosis or an alkalosis when that issue is in doubt. The test on venous blood is less expensive, and it obviates the discomfort and potential morbidity of an arterial puncture, which is required for the blood gas analysis.

We will now describe four useful tools that furnish the physician with facility in both understanding and handling acid-base problems.

Useful Tools

Conversion of pH to [H⁺]

A ready means for converting pH to $[H^+]$ allows the physician to use the Henderson equation (Eq. 3-4), since Pco_2 and $[HCO_3^-]$ will have been furnished. Although, as noted above, this computation is often provided by the laboratory, the ability to make the conversion with ease adds to the "feel" that the physician gains for the acid-base field.

Table 3-1 shows the relationship between $[H^+]$ and pH over the range of the latter that is of clinical interest. On the assumption that the activity coefficient for H^+ in blood is one — a very close approximation, given the extremely low concentration of H^+ in blood (nanomoles per liter, as opposed to millimoles per liter for the other common electrolytes, or a difference of 1,000,000) — the relationship is defined as:

$$\log [H^+] = -pH \text{ (when } [H^+] \text{ is expressed in moles/L)}; \quad (3\text{-}1)$$

for $[H^+]$ in nmoles/L (10^{-9} moles/L), the equation becomes:

$$\log [H^+] - 9 = -pH, \text{ or}$$

$$[H^+] = \text{antilog } (9 - pH) \quad (3\text{-}2)$$

The $[H^+]$ corresponding to the range of pH that is compatible with life (approximately 7 to 8) can be deduced from this equation:

Table 3-1 : True and estimated values for hydrogen ion concentration, [H$^+$], over the range of arterial plasma pH seen in patients

pH	True	Estimated*	
6.80	159	156	100/0.80, etc.
6.90	126	125	
7.00	100	100	80% of 100, etc.
7.10	79	80	
7.20	63	64	
7.30	50	50	
7.35	45	45	
7.36	44	44	
7.37	43	43	
7.38	42	42	
7.39	41	41	1 nmole/L for every 0.01 change in pH
7.40	40	40	
7.41	39	39	
7.42	38	38	
7.43	37	37	
7.44	36	36	
7.45	35	35	
7.50	32	32	
7.60	25	26	80% of 40, etc.
7.70	20	21	
7.80	16	17	
7.90	13	14	

[H$^+$] (nmoles/L)

*See text for arithmetical method of estimation.

At pH 7:

$$[H^+] = \text{antilog } (9 - 7)$$
$$= \text{antilog } 2$$
$$= 100 \text{ nmoles/L; and } -$$

At pH 8:

$$[H^+] = \text{antilog } 1$$
$$= 10 \text{ nmoles/L.}$$

Most values for pH seen in clinical practice fall between 7.00 and 7.60 (i.e., between [H$^+$] of 100 and 25 nmoles/L; Table 3-1).

The estimation of [H$^+$] from a knowledge of pH is based on two fortuitous relationships, which are apparent in Table 3-1: (1) a pH of 7.40 is equivalent to a [H$^+$] of 40 nmoles/L; and (2) in the range of pH from 7.30 to 7.45, [H$^+$] changes by 1 nmole/L for each 0.01 change in pH. Thus, over the range of pH values close to normal, the conversion is simple: For a decrease in pH, one adds 1 nmole/L for every 0.01 decrement in pH, and for increases in pH, one subtracts 1 nmole/L for every 0.01 increment in pH. For estimates outside the range pH 7.30 to 7.45, one uses the "80 rule": Starting from a pH of 7.00, which we know to be equivalent to a $[H^+]$ of 100 nmoles/L (see above), one estimates the $[H^+]$ corresponding to pH 7.10 as 80% of 100 (Table 3-1), and the $[H^+]$ for pH 7.20 as 80% of 80; for pH 7.50, one takes 80% of the $[H^+]$ at pH 7.40, which we know to be 40 nmoles/L (Table 3-1); for pH 7.60, 80% of 32, for pH 7.70, 80% of 26, for pH 7.80, 80% of 21, and for pH 7.90, 80% of 17. Finally, to estimate the $[H^+]$ for a pH below 7.00, one divides the prior value by 0.80. Thus, for pH 6.90, estimated $[H^+]$ equals 100/0.80, or 125; for pH 6.80, 125/0.80, or 156; etc. Intervening values can be interpolated.

Estimation of
[HCO₃⁻]

With the value for $[H^+]$ in hand, one can use the following form of the Henderson equation to derive $[HCO_3^-]$:

$$[H^+] = K' \frac{P_{CO_2}}{[HCO_3^-]} \tag{3-3}$$

A value for K' is derived by converting the apparent pK' for carbonic acid, 6.1, into a dissociation constant in nmoles/L (antilog 9 − 6.1) and multiplying that value by the solubility coefficient for CO_2 in plasma at 37°C, i.e., by 0.03. K' thus has a value of 794 × 0.03, or 23.8, which is rounded off to 24:

$$[H^+] = 24 \frac{P_{CO_2}}{[HCO_3^-]} \tag{3-4}$$

$$\therefore \frac{[HCO_3^-]}{(mmoles/L)} = 24 \frac{P_{CO_2} \, (mm \, Hg)}{[H^+] \, (nmoles/L)} \tag{3-5}$$

Equation 3-5 thus provides a simple arithmetic means for estimating the $[HCO_3^-]$ — "estimating," rather than "calculating," because often the value for $[H^+]$ will itself have been an approximation derived from the value for pH furnished by the laboratory. Note that the value for K' is easy to remember, since it is equal to a normal arterial $[HCO_3^-]$ of 24 mmoles/L; another way of expressing the same fact is that when P_{CO_2} and $[H^+]$ are normal, at 40 mmHg and 40 nmoles/L, respectively, $[HCO_3^-]$ is perforce equal to 24 mmoles/L.

At this point some readers must wonder why, if the modern laboratory usually supplies the $[H^+]$ and the $[HCO_3^-]$, it is necesssary to have methods for estimating these variables. There are several reasons: (1) Some laboratories still do not report these values. (2) All laboratories, no matter how good and how careful, are subject to error, and the ability to estimate values enables the physician to catch such errors. (3) The capability of deriving quick estimates at the bedside gives the physician a valuable "feel" for the subject, to which we alluded earlier. Part of that feel is to recognize ballpark values that satisfy the Henderson-Hasselbalch and Henderson relationships, and, with that recognition, to identify unusual situations or laboratory errors. The following examples illustrate the last point.

It is apparent from Equation 3-4 that whenever P_{CO_2} is equal to $[HCO_3^-]$, regardless of the value of either of these variables, $[H^+]$ will equal 24 nmoles/L, equivalent to a pH of approximately 7.60 (Table 3-1). When P_{CO_2} is twice the $[HCO_3^-]$, $[H^+]$ will equal 48 nmoles/L, or a pH of approximately 7.30; and when the P_{CO_2} is four times the $[HCO_3^-]$, $[H^+]$ approaches 100 nmoles/L, or a pH of 7.00. These general deductions thus allow one to assess quickly the approximate pH for a given set of P_{CO_2} and $[HCO_3^-]$ values. The calculation should always be made as a check for consistency, and thereby to uncover any possible error in analysis or computation or in transcription of the value to the laboratory report.

It also follows from the relationship shown in Equation 3-4 that when $[HCO_3^-]$ is low, a given change in P_{CO_2} will have a greater impact on $[H^+]$ than when $[HCO_3^-]$ is normal or high. As we shall see in following chapters, this insight has important implications for the clinical management of acid-base disorders.

Confidence Bands and Rules of Thumb. Evaluation of Compensatory Responses

In Chapter 1 (under Compensatory Responses, p. 11), we pointed out that each primary disturbance of acid-base balance is followed by a response that tends to restore the pH toward normal. The Henderson-Hasselbalch equation

$$pH = 6.1 + \log \frac{[HCO_3^-]}{0.03 \times P_{CO_2}} \tag{3-6}$$

tells us immediately the form and direction of that response. Thus, if the primary disturbance is a metabolic acidosis, which decreases $[HCO_3^-]$, then the compensatory response, *which is always in the opposite system,* must be alveolar hyperventilation, since that response will lower P_{CO_2} and thereby tend to normalize the pH (Table 3-2). Similarly, in a metabolic alkalosis, the primary change is a rise in $[HCO_3^-]$ and the compensatory response is alveolar hypoventilation, leading to an increase in P_{CO_2}.

Table 3-2 : Form and direction of compensatory responses to primary acid-base disturbances

Primary disturbance	pH	Pco2 (mm Hg)	[HCO3−] (mmoles/L)	Compensatory response
Metabolic acidosis	Low	*Low**	Low	Hyperventilation
Metabolic alkalosis	High	*High**	High	Hypoventilation
Respiratory acidosis				
Acute	Low	High	*Slightly high**	Titration of body buffers
Chronic	Slightly low	High	*High**	Increased renal reabsorption of HCO3−
Respiratory alkalosis				
Acute	High	Low	*Slightly low**	Titration of body buffers
Chronic	Slightly high	Low	*Low**	Decreased renal reabsorption of HCO3−

*The direction of the compensatory change is italicized.

Two reactions tend to restore the pH toward normal when the primary disturbance is respiratory in origin, and this fact is the basis for classifying those disturbances as acute or chronic. In respiratory acidosis there is first an immediate titration of H^+ (resulting from the processing of CO_2; Eqs. 1-1 and 1-2) by non-bicarbonate buffers (Figs. 1-2 and 1-7); during the first day, while this is the predominant compensation, the resultant state is *acute* respiratory acidosis. Thereafter, increased renal reabsorption of HCO_3^- becomes the main compensatory response (Fig. 2-2B), and after 3 to 4 days the disturbance is classified as *chronic* respiratory acidosis. The opposite changes occur during respiratory alkalosis (Table 3-2).

In order to answer Question 2 at the beginning of this chapter — i.e., Is the compensatory response quantitatively appropriate? — we need to know, in the case of metabolic disturbances, by how much the Pco2 will be lowered or raised in response to a given primary decrease or increase, respectively, in [HCO3−], and, in the case of respiratory disturbances, by how much the [HCO3−] will be raised or lowered in response to a given primary increase or decrease, respectively, in the Pco2 (Table 3-2). The answers are provided by the so-called confidence bands (Figs. 3-1 and 3-2), which form the basis for the rules of thumb (Table 3-3), to be discussed later, that are used at the bedside.

CONFIDENCE BANDS. In constructing Figure 3-1 the purpose was to ascertain the range of [HCO3−], pH, and [H+] that is seen in 95 percent of patients in whom the sole abnormality is an alteration of the Pco2 to a given level. For example, the question might be: When the Pco2 in a group of patients is raised to 60 mm Hg — and if they have no other disturbance of H+ balance — what will be the range of [HCO3−], pH, and [H+] in 95 percent of these patients? Most readers will recognize this range as being two

Primary Respiratory Disturbances

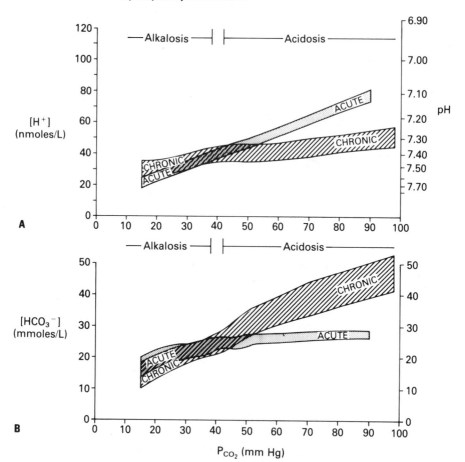

Fig. 3-1 : The 95% confidence bands for acute and chronic primary respiratory alkalosis and acidosis. *Alkalosis* is here defined as any primary respiratory disturbance in which the arterial Pco_2 is less than 38 mm Hg, even though the range of the confidence band may extend to a pH that is usually considered acidotic. Similarly, *acidosis* is here defined as any primary respiratory disturbance in which the arterial Pco_2 is greater than 42 mm Hg. Data for acute disturbances were taken from Brackett, N. C., Jr., et al., *N. Engl. J. Med.* 272:6, 1965, and Arbus, G. S., et al., *N. Engl. J. Med.* 280:117, 1969; those for chronic disturbances from Brackett, N. C., Jr., et al. *N. Engl. J. Med.* 280:124, 1969; Gennari, F. J., et al., *Clin. Res.* 28:533A, 1980; Dempsey, J. A., et al., *J. Clin. Invest.* 53:1091, 1974; Forster, H. V., et al., *J. Appl. Physiol.* 38:1067, 1975; and Severinghaus, J. W., et al., *J. Appl. Physiol.* 18:1155, 1963.

Primary Metabolic Disturbances

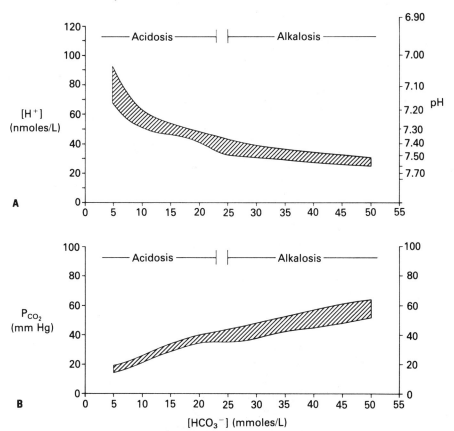

Fig. 3-2 : The 95% confidence bands for chronic metabolic acidosis and alkalosis. *Acidosis* is here defined as any primary metabolic disturbance in which the arterial [HCO₃⁻] is less than 23 mmoles per liter, even though the range of the confidence band may extend to a pH that is usually considered alkalotic. Similarly, *alkalosis* is here defined as any primary metabolic disturbance in which the arterial [HCO₃⁻] is greater than 25 mmoles per liter. Data from Winters, R. W. (ed.), *The Body Fluids in Pediatrics.* Boston: Little, Brown, 1973; van Ypersele de Strihou, C., and Frans, A., *Clin. Sci. Mol. Med.* 45:439, 1973; Bone, J. M., et al., *Clin. Sci. Mol. Med.* 46:113, 1974; and Albert, M. S., et al., *Ann. Int. Med.* 66:312, 1967.

standard deviations (S.D.) higher or lower than the mean [HCO_3^-], pH, or [H^+] at a P_{CO_2} of 60 mm Hg; or, stated differently, that any value falling outside this range will have a p-value of <0.05. The data for Figure 3-1 were obtained on normal human subjects in acute (10 minutes to 1 hour) respiratory acidosis and alkalosis, and in patients with chronic (weeks to years) respiratory acid-base disturbances. The following examples show how the bands in the figure can be used.

1. An arterial blood sample from a patient with emphysema shows a pH of 7.33 and a P_{CO_2} of 68 mm Hg. The [H^+] in this sample is 47 nmoles/L (40 + 7; Table 3-1), and the [HCO_3^-] is therefore 34.7 mmoles/L (24 × 68/47; Eq. 3-5), which is rounded off to 35 mmoles/L. As noted earlier, two mechanisms account for this elevation of HCO_3^-: (1) acutely, hydration (or hydroxylation; Eq. 1-2) of the retained CO_2 and subsequent interaction of H^+ with nonbicarbonate buffers, Buf^-:

$$CO_2 + H_2O \rightleftharpoons H_2CO_3 \rightleftharpoons H^+ + HCO_3^- \quad\quad (3-7)$$
$$\Updownarrow + Buf^-$$
$$HBuf$$

and (2) chronically, the renal response to respiratory acidosis, in which the kidneys excrete more than the normal amount of H^+ and reabsorb more HCO_3^- (Fig. 2-2B). The question that the physician needs to answer is whether in a *chronic* primary respiratory disturbance that raises the P_{CO_2} to 68 mm Hg the above mechanisms *usually* raise the [HCO_3^-] to about 35 mmoles per liter. Figure 3-1B answers this question affirmatively. The point corresponding to a P_{CO_2} of 68 mm Hg and a [HCO_3^-] of 35 mmoles per liter falls within the confidence band for chronic (but not for acute) primary respiratory acidosis; the same is true of the point corresponding to a P_{CO_2} of 68 mm Hg and a pH of 7.33 or a [H^+] of 47 nmoles per liter (Fig. 3-1A). The meaning of the points falling within the confidence bands is that in uncomplicated *chronic* respiratory acidosis, when only the two mechanisms cited above elevate the [HCO_3^-], 95 percent of the values for [HCO_3^-], pH, and [H^+] will fall within the span outlined by the bands. Thus, if the patient's history is one of chronic emphysema uncomplicated by other chronic or acute processes, the fact that the points fall within the confidence bands strengthens the physician's impression that he or she is dealing with the single, primary acid-base disturbance of chronic respiratory acidosis.

2. Suppose that an arterial blood sample from another patient with emphysema also shows the P_{CO_2} to be elevated to 68 mm Hg, but that in this patient the pH of the sample is 7.44.

By estimation, the $[H^+]$ of this sample would be 36 nmoles per liter (40 nmoles per liter minus 4 nmoles per liter), and the estimated $[HCO_3^-]$ therefore would be 45.3 mmoles/L ($24 \times 68/36$; Eq. 3-5). Locating the values for P_{CO_2} and $[HCO_3^-]$ on Figure 3-1B shows that they fall outside the confidence bands for either acute or chronic primary respiratory disturbances. This finding suggests to the physician that either a laboratory error or a mixed disturbance is involved. The choice between these alternatives can best be made on the basis of the patient's history. The patient in question was in fact admitted to the hospital, not because of the chronic emphysema but because of a peptic ulcer that had caused her to vomit for three days. Because the predominant change in H^+ balance during vomiting is due to the loss of HCl (Table 1-3), the physician knew that a mixed disturbance of metabolic alkalosis and respiratory acidosis must be present. The latter disturbance tends to elevate the $[HCO_3^-]$ by the mechanisms listed in Example 1 above; vomiting elevates the $[HCO_3^-]$ by mechanisms described in Chapter 5. The prediction in this patient is therefore that the $[HCO_3^-]$ should be elevated out of proportion to the P_{CO_2} of 68 mm Hg. This prediction is indeed fulfilled, as is reflected by the fact that the point corresponding to the arterial blood values falls above the confidence band for chronic respiratory acidosis. Reasoning another way, the physician knew before the arterial sample was drawn that a mixed disturbance of alkalosis and acidosis probably existed. Whether the arterial pH would be more alkaline or more acid than normal would depend on which of the primary disturbances predominated; in any case, it could be predicted that the pH would be more alkaline than that seen in 95 percent of patients with uncomplicated chronic emphysema and a P_{CO_2} of 68 mm Hg. This prediction is also fulfilled, as is shown by inspection of Figure 3-1A (and as indeed it must be, since the confidence bands in Figure 3-1A and those in Figure 3-1B portray the same data and are based on the Henderson-Hasselbalch equation; see also Problem 3-2).

3. When a primary metabolic disturbance is involved — either as the sole disorder or as part of a mixed disturbance, as in the patient just considered — the confidence bands of Figure 3-2 can be used. These bands are based on data from patients with primary metabolic disturbances due to various disease processes: renal failure, renal failure treated with alkali added to dialysis fluid, prolonged vomiting, excess intake of alkali, diarrhea, uncontrolled diabetes mellitus, and ingestion of NH_4Cl. In contrast to Figure 3-1, $[HCO_3^-]$ is now plotted on the abscissa because in primary metabolic disturbances $[HCO_3^-]$ is the independent variable.

Again consider the patient cited in Example 2 above, who demonstrated emphysema, peptic ulceration, and vomiting and had arterial blood sample values of pH = 7.44, P_{CO_2} = 68 mm Hg, and $[HCO_3^-]$ = 45 mmoles per liter. This patient was admitted to the hospital because of the ulcer and vomiting, not because of her emphysema. It might be expected, therefore, that attention would be focused on the metabolic alkalosis, in which there is a primary elevation of the $[HCO_3^-]$. This primary event elicits a secondary, compensatory response of alveolar hypoventilation, and the question that the confidence bands help to answer is whether the elevation of P_{CO_2} is in the range *ordinarily* to be expected in uncomplicated, primary metabolic alkalosis. For the patient being considered, the P_{CO_2} of 68 mm Hg is higher than what occurs in 95 percent of patients or normal subjects in whom a primary metabolic alkalosis has caused the $[HCO_3^-]$ to rise to 45 mmoles per liter. Thus, the patient had hypoventilated more than would be expected as a compensatory response to this degree of primary metabolic alkalosis, the reason being that the patient had, in addition, a primary cause for alveolar hypoventilation, namely, chronic emphysema.

Note: The use and interpretation of the confidence bands is not always so straightforward as in the preceding discussion. For example, in the alkalotic range for respiratory disturbances (Fig. 3-1), the bands for acute and chronic disorders overlap to such an extent that separation of the two entities solely from confidence bands is difficult or impossible (an example is given in Chap. 7, under Hepatic Failure). In addition, W. B. Schwartz and his associates have rightly emphasized that while values that fall outside of the bands denote a mixed disturbance, *points falling within the bands do not necessarily reflect a single uncomplicated, primary disturbance of H^+ balance.* The following clinical example will illustrate this point (see also Chap. 8, under Acute Pulmonary Decompensation in Emphysema).

4. A middle-aged woman with chronic obstructive pulmonary disease has right-sided heart failure and edema. She has been treated with furosemide, which led to selective loss of Cl^- and to metabolic alkalosis (see Chap. 5), so that her $[HCO_3^-]$ rose from 31 to 40 mmoles per liter. At this point, her arterial pH is 7.45; her P_{CO_2}, 58 mm Hg. We know from the history that these laboratory values must reflect a mixed disturbance of chronic respiratory acidosis and metabolic alkalosis, and in fact the values fall outside the confidence bands in Figure 3-1. The patient then developed an acute respiratory infection, which further embarrassed her ventilation, and she was hospitalized. On admission, a blood gas

analysis revealed the following: pH = 7.37, P_{CO_2} = 75 mm Hg, and $[HCO_3^-]$ = 42 mmoles/L. These values fall within the confidence band for chronic respiratory acidosis in Figure 3-1. At this point, therefore, the "automatic" usage of confidence bands (or rules of thumb; see below) might lead the physician to conclude, incorrectly, that the patient has an uncomplicated chronic respiratory acidosis, whereas she really has a mixed disturbance of acute and chronic respiratory acidosis together with metabolic alkalosis. This example emphasizes the point made earlier, namely, that automatic usage of aids or graphs can lead to erroneous conclusions, and that *laboratory data should always be interpreted in light of the patient's history.* An extension of this maxim is that one needs to use clinical judgment to determine which of the bands to use (i.e., the one for primary respiratory or for primary metabolic disturbances).

RULES OF THUMB. Because it is awkward to work with graphs, such as Figures 3-1 and 3-2, at the bedside, the rules of thumb in Table 3-3 have been developed. These guidelines were derived from the slopes depicted in Figures 3-1 and 3-2, but the rules of thumb define only a narrow mean and not the broad variance of the confidence bands. Consequently, these rules, while convenient, have certain limitations: The response can be considered

Table 3-3 : Rules of thumb for gauging compensatory responses at the bedside

Primary disturbance	Normal compensatory response
Metabolic acidosis	For each mmole/L fall in $[HCO_3^-]$, P_{CO_2} decreases by 1.3 mm Hg
Metabolic alkalosis	For each mmole/L rise in $[HCO_3^-]$, P_{CO_2} increases by 0.7 mm Hg
Respiratory acidosis	
Acute	For each mm Hg rise in P_{CO_2}: $[HCO_3^-]$ increases by 0.1 mmole/L $[H^+]$ increases by 0.7 nmole/L
Chronic	For each mm Hg rise in P_{CO_2}: $[HCO_3^-]$ increases by 0.4 mmole/L $[H^+]$ increases by 0.3 nmole/L
Respiratory alkalosis	
Acute	For each mm Hg fall in P_{CO_2}: $[HCO_3^-]$ decreases by 0.2 mmole/L $[H^+]$ decreases by 0.7 nmole/L
Chronic	For each mm Hg fall in P_{CO_2}: $[HCO_3^-]$ decreases by 0.4 mmole/L $[H^+]$ decreases by 0.4 nmole/L

abnormal only when it deviates quite widely from the rule being tested. Fortunately, in most instances of an abnormal response, the deviation is sufficiently marked to permit an unambiguous conclusion.

The utility of the rules of thumb can be illustrated by returning to the clinical histories discussed above. In the case of Patient 1, the history of emphysema suggests that we are dealing with chronic respiratory acidosis. In that condition, according to the rules of thumb (Table 3-3), the $[HCO_3^-]$ should rise by approximately 0.4 mmole/L for each mm Hg increment in PCO_2. In the patient, PCO_2 rose by 28 mm Hg (68 − 40). Multiplying that value by 0.4 yields 11.2, or an 11 mmoles/L rise in $[HCO_3^-]$; and adding that amount to the normal concentration of 24 mmoles/L gives a new pre-dicted plasma $[HCO_3^-]$ of 35 mmoles/L. This figure is identical to the computed value of 35 mmoles/L. Thus, the laboratory re-ports confirm the clinical impression of uncomplicated chronic respiratory acidosis.

Because the relationship between PCO_2 and $[HCO_3^-]$ tends to be curvilinear in primary respiratory disturbances (Fig. 3-1), while that between PCO_2 and pH or $[H^+]$ is more or less straight, some physicians prefer rules of thumb that predict values for pH and $[H^+]$ (Table 3-3). For Patient 1, with chronic respiratory acidosis and a rise in PCO_2 of 28 mm Hg, the rules would thus predict a rise in $[H^+]$ of 8 nmoles/L (28 × 0.3), or a new value of 48 nmoles/L, which corresponds to a pH of 7.32 (7.40 − 0.08; Table 3-1).

In the case of Patient 2, the history of peptic ulceration and vomiting that precipitated her admission to the hospital should focus attention on metabolic alkalosis. The $[HCO_3^-]$ in this pa-tient had risen to 45 mmoles/L, i.e., by 21 mmoles/L (45 − 24). The rules of thumb for metabolic alkalosis (Table 3-3) therefore predict a new PCO_2 of 55 mm Hg [40 + (21 × 0.7)], which is considerably lower than the reported value of 68 mm Hg. The calculation thus confirms the physician's impression that there is an additional disturbance producing further elevation of the PCO_2, and that disorder is the emphysema.

Alternatively, if one considered emphysema to be the prime disorder in this patient, the rules of thumb for chronic respira-tory acidosis would predict an increase in $[HCO_3^-]$ to 35 mmoles/L [24 + (28 × 0.4)], and a new $[H^+]$ of 48 nmoles/L [40 + (28 × 0.3)], or a pH of 7.32. All these values lead to the same conclusion: that the patient must have an additional acid-base disturbance (namely, metabolic alkalosis from vomiting), which has raised her plasma $[HCO_3^-]$ far above the predicted value and has, in fact, resulted in a final state of alkalosis, with a $[H^+]$ of 36 nmoles/L and a pH of 7.44.

Finally, in Patient 4, before she developed the acute respiratory infection, a compensatory rise in $[HCO_3^-]$ to 31 mmoles/L $[24 + (18 \times 0.4)]$ would have been predicted on the basis of chronic respiratory acidosis (Table 3-3). The fact that the measured $[HCO_3^-]$ was 40 mmoles/L signaled the presence of a mixed component, namely, metabolic alkalosis due to diuretics, which raised the $[HCO_3^-]$ out of the expected range. When this patient developed a respiratory infection, an acute respiratory acidosis was superimposed on the chronic respiratory acidosis, and therefore it became impossible to know which correction factor to use to predict the expected compensatory response (Table 3-3). Clinical judgment based on the patient's history, however, left no doubt that the patient had a mixture of acute and chronic respiratory acidoses plus metabolic alkalosis, and it was this judgment, not solely laboratory data, that determined how the patient's illness was to be managed.

Anion Gap

This variable is helpful in answering the third question posed at the beginning of this chapter — namely, What is the cause of a given acid-base disturbance? — especially as the question pertains to the differential diagnosis of metabolic acidoses (Table 4-1).

DEFINITION. Figure 3-3 shows that most of the electrical charges on the major cation of plasma — namely, Na^+ — are neutralized by Cl^- and HCO_3^-. The remaining charges on Na^+, which amount to 8 to 16 meq per liter, are covered by some of the negative charges on the other anions in plasma, namely, phosphate, sulfate, organic acids, and proteins. The sum of these charges is called the anion "gap" because they cover the gap of positive charges on Na^+ left, so to speak, by Cl^- and HCO_3^-. The *anion gap* is thus defined as the difference between $[Na^+]$ and the sum of $[Cl^-]$ and $[HCO_3^-]$, all expressed in milliequivalents per liter (meq/L); or,

$$\text{Anion gap} = [Na^+] - ([Cl^-] + [HCO_3^-]) \qquad (3\text{-}8)$$

If one considers the range of normal concentrations for Na^+, Cl^-, and HCO_3^- listed in Table 1-1, one can derive a *normal range for the anion gap of 8 to 16 meq per liter.* The wide range results mainly from the fact that the calculation of the gap includes normal analytic variations of three separate laboratory determinations. An unequivocal increase in the anion gap is present if the value exceeds 19 or 20 meq per liter. A reduction in the gap is rarely seen.

The gap is sometimes referred to as the "sum of unmeasured anions" because phosphate, sulfate, organic acids, and proteins are not so *routinely* measured as are Cl^- and HCO_3^-. In fact, in

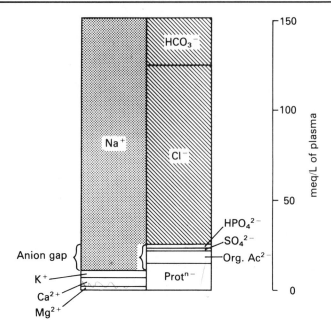

Fig. 3-3 : Anion gap for normal plasma.

modern clinical jargon, when a physician refers to "serum electrolytes" he or she means Na^+, K^+, Cl^-, and HCO_3^-, leaving out the other electrolytes in plasma (Fig. 3-3). Thus, although "sum of unmeasured anions" is a correct description insofar as these anions are not often measured, it is a misnomer because the anion gap does not refer to the total charges on these anions but only to the equivalents that are required to neutralize the remaining charges on Na^+.

MEANING OF ANION GAP. Table 3-4 lists examples of the value of the anion gap in various diseases. *The major clinical usefulness of the anion gap lies in the differential diagnosis of metabolic acidosis:* this fact is reflected in Table 4-1.

An increased anion gap is seen in some forms of metabolic acidosis that are due, at least in part, to an accumulation of organic acids. Perhaps the prototype is uncontrolled diabetes mellitus, in which 3-hydroxybutyric acid and acetoacetic acid are produced at a rate faster than they can be metabolized or excreted. With a pK' of 4.8 and 3.8, respectively (Table 1-2), these acids give up H^+ (protons) at the pH of body fluids (for a dynamic visualization of this process, see Fig. 1-5); HCO_3^- is consumed as the H^+ is buffered. Thus, for each mmole of these acids added to extracellular fluid, 1 mmole of HCO_3^- is consumed (as in Eq. 1-11). As a result, plasma $[HCO_3^-]$ falls in direct proportion to the accumulation of these acids, and it is replaced

Table 3-4 : Values for the anion gap in several disease states. Normal anion gap = 8 to 16 meq/L

State	Arterial pH	Venous plasma concentration (mmoles/L)			Anion gap (meq/L)	Comments
		Na^+	Cl^-	HCO_3^-		
Uncontrolled diabetes mellitus	7.10	138	101	5	32	Unmeasured anions are mainly 3-hydroxybutyrate and acetoacetate
Spontaneous lactic acidosis	7.10	137	98	10	29	Unmeasured anion is lactate, which in this patient was 17 meq/L
Chronic renal failure	7.32	138	102	18	18	Unmeasured anions include those of organic acids, sulfate, and phosphate
Prolonged vomiting	7.58	141	82	40	19	Unmeasured anions are mainly those of ketoacids produced during increased fat metabolism (see Fig. 4-1). (Note: An increased anion gap does not necessarily mean that the patient has a low pH.)
Anxiety-hyperventilation syndrome	7.58	137	103	22	12	Unmeasured anions do not increase, or increase only slightly, in acute respiratory alkalosis because the slight fall in HCO_3^- is counterbalanced by a shift of Na^+ into cells and, hence, a fall in plasma Na^+ concentration
Chronic hypoxic hyperventilation	7.49	139	108	15	16	This condition is accompanied by very slight accumulation of unmeasured anions, which accounts for the anion gap being at the upper limit of normal. The nature of these anions is not known
Ingestion of NH_4Cl	7.12	143	122	7	14	Note: Metabolic acidosis can occur without an increase in anion gap (for reaction, see text)
Diarrhea and other HCO_3^--losing states	7.32	148	116	20	12	(Same comment as above)
Chronic obstructive lung disease	7.37	142	95	35	12	Uncomplicated chronic respiratory acidosis is not accompanied by an increase in the anion gap

by the anions of the acids — in this instance, by 3-OH butyrate and acetoacetate. Na^+, however, often remains at a normal concentration, as does Cl^-. It follows from Equation 3-8 that the anion gap must be increased, a point that can also be understood from Figure 3-3 if one visualizes that the major change in the plasma profile is an increase in organic acids at the expense of HCO_3^-.

Another example is lactic acidosis, a condition in which the concentration of this organic acid, which is the main end product of anaerobic metabolism, may rise from its normal value of about 1 meq per liter to 10 or 20 meq per liter or higher. Again, the H^+ that is added to the extracellular fluid by the dissociation of lactic acid (pK' = 3.9) is buffered by HCO_3^-, thereby decreasing the plasma concentration of HCO_3^- and increasing that of the anion, lactate. Lactic acidosis occurs in association with a variety of disorders, which are listed in Table 4-5.

Chronic renal failure is a common cause of an elevated anion gap. In this condition a number of anions, such as sulfate, phosphate, and urate, accumulate in the plasma. The H^+ that is associated with these anions is largely buffered by HCO_3^- and renal failure is therefore accompanied by a metabolic acidosis characterized by an increased anion gap.

It should be noted that an increase in the anion gap is not always associated with a pH that is less than normal. For example, after heavy intake of alcohol, severe and prolonged vomiting may ensue. This development initially induces a large loss of HCl from the body, and hence leads to alkalosis. If vomiting persists, however, hypotension, relative tissue ischemia, and consequent lactic acidosis can develop. As long as the loss of HCl exceeds the accumulation of lactic acid, the resultant arterial pH will remain alkaline (>7.40). Nonetheless, the footprints of augmented lactic acid production will be an increase in the anion gap. In other words, a major increase in this gap is prima facie evidence that one component of a mixed disturbance is metabolic acidosis, even if the pH happens to be alkaline.

Conversely, acidosis is not necessarily accompanied by a rise in the anion gap (Table 4-1). Ingestion of NH_4Cl causes a metabolic acidosis through the net addition of HCl, as in the following reaction:

$$2 NH_4Cl + CO_2 \rightarrow 2 HCl + H_2O + CO(NH_2)_2$$

Most of the H^+ produced in this reaction is buffered by extracellular HCO_3^-, thereby reducing the $[HCO_3^-]$, but for every equivalent of HCO_3^- thus taken out of solution, an equivalent of Cl^- (from the HCl) is added; hence, the anion gap does not change

(Eq. 3-8 and Fig. 3-3). Further examples include diarrhea or loss of other gastrointestinal fluids with high concentrations of HCO_3^-, where again Cl^- "replaces" HCO_3^-. An analogous chain of events keeps the anion gap normal in uncomplicated respiratory acidosis, as in chronic obstructive lung disease. The CO_2 that is retained in this condition is hydrated (or hydroxylated; see Eq. 1-2), as shown in Equation 3-7. The H^+ thus formed has to be buffered almost exclusively within cells (i.e., by Buf^-), because the main extracellular buffer, HCO_3^-, cannot be used to neutralize this H^+. (This fact is apparent from Eq. 3-7; the reaction would have to shift to the left if HCO_3^- were to buffer the H^+, but it is already being shifted to the right by the primary addition of CO_2.) The HCO_3^- produced in this reaction stays mainly within the extracellular space. According to Equation 3-8, this addition might be expected to reduce the anion gap; no reduction occurs, however, because the rise in $[HCO_3^-]$ is accompanied by a reciprocal decrease in $[Cl^-]$. The reason for the reciprocity probably involves the prime importance of Na^+ balance; that is, Na^+ reabsorption continues unabated, and since plasma $[HCO_3^-]$ and hence the filtered load of HCO_3^- (GFR $\times$ $P_{HCO_3^-}$) are increased, HCO_3^- preferentially accompanies the reabsorbed Na^+ while the reabsorption of Cl^- is reduced. In fact, Cl^- excretion is increased during adaptation to chronic respiratory acidosis.

Finally, calculation of the anion gap can provide useful supplemental information in other acid-base disturbances. For example, in acute respiratory alkalosis (as during mechanical or psychogenic hyperventilation), the gap should be normal; the reason is that the small decline in bicarbonate concentration (consequent upon the reaction in Eq. 3-7 shifting to the left) is accompanied by a proportional decline in the Na^+ concentration. In chronic respiratory alkalosis, however, the anion gap is sometimes raised slightly above the upper limit of normal (by perhaps 3 meq/L). The explanation for the increase is not clear; it is not due to accumulation of lactate. In uncomplicated metabolic alkalosis there may be a slight increase in the anion gap; it rarely exceeds 4 or 5 meq/L, although it can be higher if the alkalosis is very severe (plasma $[HCO_3^-] > 50$ mmoles/L). Not all the explanations for the increase are known. A portion of the rise is the result of an increase in the free anionic groups of protein buffers. As the arterial pH rises, the titration of these buffers releases H^+ from them, thereby raising the number of negative charges on the proteins (Fig. 1-5).

Summary

The laboratory analysis of clinical problems in acid-base balance is based on the Henderson-Hasselbalch equation (Eq. 3-6). This fact often constitutes a serious stumbling block to the understanding and quantitative management of acid-base disorders, because many physicians are not conversant with manipulation of the logarithms in this equation. There are two general solutions to this problem: (1) Many laboratories now automatically compute and report all three variables of the Henderson-Hasselbalch equation, so that logarithmic calculations are obviated. (2) One can use a modification of the Henderson equation (Eq. 3-4), which displays the same relationship as the Henderson-Hasselbalch equation, but in arithmetic form. This equation allows rapid manipulation of the variables of interest, and provides a basis for developing a first-hand dexterity in evaluating acid-base disturbances. Two useful tools that facilitate the manipulation are the conversion of pH to $[H^+]$ and the estimation of $[HCO_3^-]$.

Two additional tools have been presented. One is the use of rules of thumb (Table 3-3) to determine whether a simple or mixed disturbance of acid-base balance is present. In the case of primary metabolic disturbances, the rules allow the physician to determine whether the compensatory change in P_{CO_2} is of the expected magnitude, and, in the case of primary respiratory disturbances, whether the compensatory change in plasma $[HCO_3^-]$ is quantitatively appropriate. Should the patient's laboratory values fall outside the expected response, a mixed disturbance is indicated (provided that a laboratory error has been excluded); if, however, the values fall within the expected response, this fact will not *necessarily* reflect an uncomplicated primary disturbance.

The final tool discussed in this chapter is the anion gap, which is calculated as the difference between the plasma $[Na^+]$ and the sum of the plasma $[Cl^-]$ and $[HCO_3^-]$. Because the gap is typically increased or unchanged in certain acid-base disturbances, its quick calculation at the bedside is often of tremendous help in analyzing such disturbances, especially different types of metabolic acidosis (Table 4-1).

In the next five chapters we will illustrate how these tools can be utilized in the management of patients with acid-base disturbances. It has been repeatedly emphasized that the correct analysis of such disturbances, and hence the logical management of such patients, requires an understanding of the dynamics of the disorder. For this reason — and for the reason that a clinical diagnosis should be based on a history and physical examination *supported* by laboratory studies, seldom on laboratory values alone — "automatic" diagrams of H^+ imbalances have not been presented.

Problem 3-1   Under the following columns are listed variables of the Henderson-Hasselbalch equation or the Henderson equation for components of arterial plasma. In each instance, fill in the blank spaces (not occupied by dashes) by supplying both the calculated (Calc.) and the estimated (Est.) values. Also supply the units in the parentheses.

	pH		$[H^+]$ ( )		Pco_2 ( )		$[HCO_3^-]$ ( )	
	Calc.	Est.	Calc.	Est.	Calc.	Est.	Calc.	Est.
(1)	7.40	—			40	—	24	—
(2)	7.32	—	—	—			15	—
(3)	7.55	—	—	—	44	—		
(4)			65	—	75	—	28	—

Problem 3-2

A 61-year-old woman was transferred from another hospital because she was agitated and hallucinating. She had been admitted to the previous hospital 10 days earlier because of nausea and vomiting, and while there she had become increasingly confused. Her most striking laboratory finding at that hospital had been a plasma $[HCO_3^-]$ of 50 mmoles per liter.

Some 3½ years before the present admission she had consulted a physician because of a urinary tract infection. At that time an ultrasound examination revealed that both her kidneys were small. Her serum creatinine concentration had been 3.8 mg/100 ml; the concentration of urea nitrogen in her blood (BUN), 90 mg/100 ml. These findings were interpreted to reflect chronic renal insufficiency. Several determinations of her plasma $[Ca^{2+}]$ had been borderline high, but this finding was not satisfactorily explained or pursued.

Physical examination on admission revealed a small, thin woman with dry skin and dry mucous membranes. She was hallucinating, and frequently broke into tears.

Because of her mental status (for which there was no immediate explanation) and because the main laboratory abnormality re-

Table 3-5 : Results of laboratory tests on a 61-year-old woman on admission to the hospital

Test	On Admission
Arterial blood:	
pH	7.51
Pco_2 (mm Hg)	34
Venous serum:	
Total CO_2 (mmoles/L)	45
$[Cl^-]$ (mmoles/L)	98
$[Na^+]$ (mmoles/L)	152
BUN (mg/100 ml)	82
Creatinine (mg/100 ml)	3.1

ported from the other hospital was a very high plasma [HCO$_3^-$], the studies listed in Table 3-5 were obtained shortly after the patient was admitted.

Analyze the patient's acid-base status. What would you have done next?

Appendix: Nomogram of the Henderson-Hasselbalch Equation

The Henderson-Hasselbalch equation has three variables: pH, [HCO$_3^-$], and P$_{CO_2}$; that is,

$$pH = 6.1 + \log \frac{[HCO_3^-]}{0.03 \times P_{CO_2}}$$

When any two of these variables are known, the third can be derived from the nomogram of Figure 3-4 by laying a straight-edge to intersect the two known variables. For example, if the laboratory reports that a patient's arterial blood, taken anaerobically, has a pH of 7.32 and a P$_{CO_2}$ of 32 mm Hg, then the nomogram tells us that the [HCO$_3^-$] in that same arterial sample must be 16 mmoles per liter. Although computers and most calculators readily make the calculation, and even though the laboratory often automatically reports all three variables, the nomogram nevertheless continues to be a useful tool when the reader wants to make a quick calculation or to check for consistency of data.

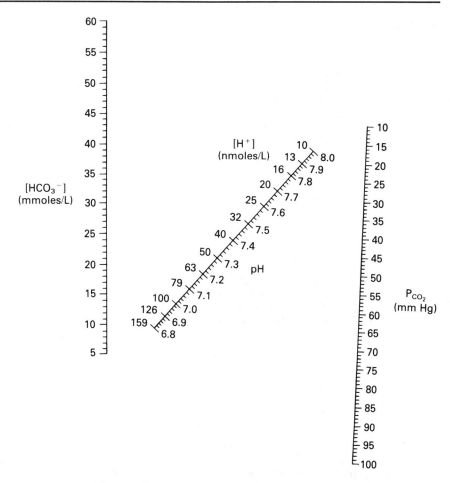

Fig. 3-4 : Nomogram of the Henderson-Hasselbalch equation. Each variable extends over the ranges shown in Figs. 3-1 and 3-2, or slightly beyond. Adapted from McLean, F. C., *Physiol. Rev.* 18:495, 1938; Davenport, H. W., *The ABC of Acid-Base Chemistry* (6th ed.). Chicago: University of Chicago Press, 1974.

**Selected
References**

General

Cohen, J. J., and Kassirer, J. P. *Acid-Base.* Boston: Little, Brown, 1982. Chap. 7.

Davenport, H. W. *The ABC of Acid-Base Chemistry* (6th ed.). Chicago: University of Chicago Press, 1974.

Davis, R. P. Logland: A Gibbsian view of acid-base balance. *Am. J. Med.* 42:159, 1967.

Hills, A. G. *Acid-Base Balance: Chemistry, Physiology, Pathophysiology.* Baltimore: Williams & Wilkins, 1973.

Huckabee, W. E. Henderson vs. Hasselbalch. *Clin. Res.* 9:116, 1961.

Masoro, E. J., and Siegel, P. D. *Acid-Base Regulation: Its Physiology, Pathophysiology, and Interpretation of Blood-Gas Analysis* (2nd ed.). Philadelphia: Saunders, 1977.

Nahas, G. G. (ed.). Current concepts of acid-base measurement. *Ann. N.Y. Acad. Sci.* 133:1, 1966.
Two of the papers in this symposium, Terminology of Acid-Base Disorders and Statement on Acid-Base Terminology, have also been reproduced in Ann. Intern. Med. *63:873 and 885, 1965.*

Narins, R. G., and Emmett, M. Simple and mixed acid-base disorders: A practical approach. *Medicine* 59:161, 1980.

Rector, F. C., Jr. (ed.). Symposium on acid-base homeostasis. *Kidney Int.* 1:273, 1972.

Rose, B. D. *Clinical Physiology of Acid-Base and Electrolyte Disorders* (2nd ed.). New York: McGraw-Hill, 1984. Chap. 17.

Schwartz, W. B., and Relman, A. S. Critique of the parameters used in evaluation of acid-base disorders. "Whole-blood buffer base" and "standard bicarbonate" compared with blood pH and plasma bicarbonate concentration. *N. Engl. J. Med.* 268:1382, 1963.

Valtin, H. *Renal Dysfunction: Mechanisms Involved in Fluid and Solute Imbalance.* Boston: Little, Brown, 1979. Chaps. 5 and 6.

Weisberg, H. F. Water, Electrolytes, Acid-Base, and Oxygen. In I. Davidson and J. B. Henry (eds.), *Todd-Sanford Clinical Diagnosis by Laboratory Methods* (15th ed.). Philadelphia: Saunders, 1974.

Winters, R. W. (ed.). *The Body Fluids in Pediatrics.* Boston: Little, Brown, 1973.

*Arithmetic
Approximations*

Fagan, T. J. Estimation of hydrogen ion concentration. *N. Engl. J. Med.* 288:915, 1973.

Flenley, D. C. Another non-logarithmic acid-base diagram? *Lancet* 1:961, 1971.

Kassirer, J. P., and Bleich, H. L. Rapid estimation of plasma carbon dioxide tension from pH and total carbon dioxide content. *N. Engl. J. Med.* 272:1067, 1965.

Confidence Bands   Albert, M. S., Dell, R. B., and Winters, R. W. Quantitative displacement of acid-base equilibrium in metabolic acidosis. *Ann. Intern. Med.* 66:312, 1967.

Arbus, G. S., Herbert, L. A., Levesque, P. R., Etsten, B. E., and Schwartz, W. B. Characterization and clinical application of the "significance band" for acute respiratory alkalosis. *N. Engl. J. Med.* 280:117, 1969.

Bone, J. M., Cowie, J., Lambie, A. T., and Robson, J. S. The relationship between arterial Pco_2 and hydrogen ion concentration in chronic metabolic acidosis and alkalosis. *Clin. Sci. Mol. Med.* 46:113, 1974.

Brackett, N. C., Jr., Cohen, J. J., and Schwartz, W. B. Carbon dioxide titration curve of normal man: Effect of increasing degrees of acute hypercapnia on acid-base equilibrium. *N. Engl. J. Med.* 272:6, 1965.

Brackett, N. C., Jr., Wingo, C. F., Muren, O., and Solano, J. T. Acid-base response to chronic hypercapnia in man. *N. Engl. J. Med.* 280:124, 1969.

Bushinsky, D. A., Coe, F. L., Katzenberg, J., Szidon, J. P., and Parks, J. H. Arterial Pco_2 in chronic metabolic acidosis. *Kidney Int.* 22:311, 1982.

Cohen, J. J., and Schwartz, W. B. Evaluation of acid-base equilibrium in pulmonary insufficiency. *Am. J. Med.* 41:163, 1966.

Dempsey, J. A., Forster, H. V., and DoPico, G. A. Ventilatory acclimatization to moderate hypoxemia in man. The role of spinal fluid [H^+]. *J. Clin. Invest.* 53:1091, 1974.

Forster, H. V., Dempsey, J. A., and Chosy, L. W. Incomplete compensation of CSF [H^+] in man during acclimatization to high altitude (4,300 m). *J. Appl. Physiol.* 38:1067, 1975.

Gennari, F. J., Goldstein, M. B., and Schwartz, W. B. The nature of the renal adaptation to chronic hypocapnia. *J. Clin. Invest.* 51:1722, 1972.

Lennon, E. J., and Lemann, J., Jr. Defense of hydrogen ion concentration in chronic metabolic acidosis. A new evaluation of an old approach. *Ann. Intern. Med.* 65:265, 1966.

Madias, N. E., Bossert, W. H., and Adrogué, H. J. Ventilatory response to chronic metabolic acidosis and alkalosis in the dog. *J. Appl. Physiol: Respiratory, Environmental and Exercise Physiol.* 56:1640, 1984.

Schwartz, W. B., Brackett, N. C., Jr., and Cohen, J. J. The response of extracellular hydrogen ion concentration to graded degrees of chronic hypercapnia: The physiologic limits of the defense of pH. *J. Clin. Invest.* 44:291, 1965.

Severinghaus, J. W., Mitchell, R. A., Richardson, B. W., and Singer, M. M. Respiratory control at high altitude suggesting active transport regulation of CSF pH. *J. Appl. Physiol.* 18:1155, 1963.

van Ypersele de Strihou, C., and Frans, A. The respiratory response to chronic metabolic alkalosis and acidosis in disease. *Clin. Sci. Mol. Med.* 45:439, 1973.

Anion Gap   Adrogué, H. J., Brensilver, J., and Madias, N. E. Changes in the plasma anion gap during chronic metabolic acid-base disturbances. *Am. J. Physiol.* 235 (Renal Fluid Electrolyte Physiol. 4):F291, 1978.

DeTroyer, A., Stolarczyk, A., Zegers DeBeyl, D., and Stryckmans, P. Value of anion-gap determination in multiple myeloma. *N. Engl. J. Med.* 296:858, 1977.

Emmett, M. E., and Narins, R. G. Clinical use of the anion gap. *Medicine* (Baltimore) 56:38, 1977.

Gabow, P. A. Disorders associated with an altered anion gap. *Kidney Int.* 27:472, 1985.

Oh, M. D., and Carroll, H. J. The anion gap. *N. Engl. J. Med.* 297:814, 1977.

4 : Metabolic Acidosis: Clinical Examples

In this and the following chapters, our approach to problems of H^+ imbalance will be illustrated by means of clinical presentations. Examples of each of the primary disturbances will be considered in Chapters 4 through 7, which will include tables listing the major disease processes that can lead to a particular acid-base disorder. Examples of common mixed acid-base disturbances will be discussed in the final chapter.

Causes

The causes of metabolic acidosis are listed in Table 4-1. Note that the conditions are divided on the basis of whether they are characterized by a normal or an increased anion gap.

Table 4-1 : Causes of metabolic acidosis
The disorders have been divided into those that are accompanied by an increased anion gap and those accompanied by a normal anion gap.

Increased Anion Gap:
    Uncontrolled diabetes mellitus
    Lactic acidosis
    Renal failure (acute and chronic)
    Administration, ingestion, or intoxication:
        Ethyl alcohol, with "starvation" and production of keto acids
        Salicylate
        Methyl alcohol
        Paraldehyde
        Ethylene glycol

Normal Anion Gap:
    Diarrhea or loss of other gastrointestinal fluids with high $[HCO_3^-]$
        through fistulas or surgical drainage
    Renal tubular acidosis (RTA)
        Proximal RTA
        Distal RTA
    Ureterosigmoidostomy
    Administration, ingestion, or intoxication:
        NH_4Cl
        Carbonic anhydrase inhibitors

Chronic Renal Failure

A 32-year-old man had been followed for seven years by his physician because of renal insufficiency. At age 20 he had had a sore throat due to β-hemolytic Streptococcus, group A. Two to three weeks after the onset of the sore throat, he suddenly began to pass bright red (bloody) urine. On evaluation by his physician, he was found to be slightly hypertensive, and analysis of his urine revealed a moderate amount of protein and erythrocytes, as well as red blood cell casts in the sediment. A diagnosis of acute poststreptococcal glomerulonephritis was made.

Thereafter, the patient was checked by his physician every three to six months. He continued to excrete excess protein in his urine. Twelve years after the episode of sore throat, the patient was feeling well except that he seemed to tire somewhat easily. He was working full-time and following a regular diet. Physical examination was unremarkable except for borderline hypertension (BP 150/90 mm Hg). The results of laboratory tests obtained at this time are listed in Table 4-2.

Comment

The history suggests renal insufficiency due to chronic glomerulonephritis. This impression is supported by the elevations in serum creatinine and BUN, which reflect a decrease in the glomerular filtration rate (GFR).

In chronic renal insufficiency we expect to find metabolic acidosis, caused mainly by a failure of renal excretion of acid to keep pace with the daily production of fixed acids from dietary precursors. Typically, as in the present patient, the anion gap is increased (Tables 4-1 and 4-2) because the renal excretion of phosphate, sulfate, and organic acids (e.g., uric acid) is insufficient to maintain balance for these anions. (When renal insufficiency is mild, the anion gap may be normal because at this

**Table 4-2 : Results of laboratory tests
in a 32-year-old man with chronic renal failure**
For $[H^+]$ and $[HCO_3^-]$, the first figure is the calculated value, the one in parentheses the estimated value.

Test	Result
Arterial blood:	
pH	7.34
P_{CO_2} (mm Hg)	32
$[H^+]$ (nmoles/L)	46 (46)
$[HCO_3^-]$ (mmoles/L)	17 (17)
Venous serum:	
$[Na^+]$ (mmoles/L)	142
$[Cl^-]$ (mmoles/L)	102
$[K^+]$ (mmoles/L)	4.5
Total CO_2 (mmoles/L)	18
Anion gap (meq/L)	22*
Creatinine (mg/100 ml)	9.8
Blood urea nitrogen (BUN) (mg/100 ml)	95

*Computed on the basis of venous total CO_2, as is the usual practice. Using the arterial $[HCO_3^-]$ might be more accurate, but the quantitative differences between the two computations is not important (see Chap. 3, under Venous Total CO_2, p. 62).

stage the renal excretion of the above-named anions can still keep pace with their production.)

If we are dealing with the uncomplicated metabolic acidosis of chronic renal failure (uremic acidosis), we would expect the fall in arterial P_{CO_2} to follow the rule of thumb for metabolic acidosis (Table 3-3), and the plasma acid-base values to fall within the confidence bands given in Figure 3-2. In addition, we would anticipate the anion gap to be increased above normal.

These predictions are all fulfilled. By calculation or estimation (Table 3-1), the $[H^+]$, corresponding to a pH of 7.34, is 40 + 6, or 46 nmoles per liter, and the $[HCO_3^-]$ is 24 × 32/46 (Eq. 3-5), or 17 mmoles per liter. The value for $[HCO_3^-]$, calculated by the Henderson-Hasselbalch equation (Eq. 3-6), is also 17 mmoles per liter. When we use the rule of thumb for metabolic acidosis (Table 3-3), arterial P_{CO_2} is expected to fall by 1.3 mm Hg (on average) for each mmole per liter decrement in $[HCO_3^-]$. In this case, $[HCO_3^-]$ has decreased by 7 mmoles per liter (from the average normal value of 24 to 17), and therefore arterial P_{CO_2} would be expected to decrease by 1.3 × 7, or 9 mm Hg. When we use the average normal value for arterial P_{CO_2} of 40 mm Hg, the measured P_{CO_2} has fallen by 8 mm Hg. This change is close enough to that predicted by the rule of thumb to be consistent with the increase in ventilation expected in an otherwise normal subject with this degree of metabolic acidosis. Similarly, if one uses the confidence bands illustrated in Figure 3-2, the arterial values fall within the bands for uncomplicated primary metabolic acidosis.

The anion gap (Eq. 3-8) is usually calculated using the values obtained on venous serum or plasma because sodium and chloride, as well as total CO_2 content, are routinely measured on the same sample. In this instance, the anion gap, $Na^+ - ([Cl^-] + total\ CO_2)$, is 22 meq per liter. Venous total CO_2 is usually 1 to 3 mmoles per liter higher than arterial bicarbonate concentration (see Chap. 3, under Laboratory Tests, p. 62), partly because it includes dissolved CO_2 and partly because venous bicarbonate concentration is slightly higher than arterial bicarbonate concentration. However, this is a trivial distinction, for whether one uses 17, 18, or even 19 mmoles per liter for the $[HCO_3^-]$ value, the anion gap is clearly elevated in this patient.

The laboratory values thus confirm the diagnostic impression based on the clinical history. It should be noted that a patient with chronic renal insufficiency — even, as this patient, with a reduction in GFR to less than 15 percent of normal — can remain in balance for Na^+, Cl^-, and K^+ while ingesting a normal diet. Not so, however, in the case of H^+, for which there is a positive balance; these ions are thought to be buffered mainly by bone, so that the arterial pH may remain stable despite continual retention of acid as renal failure advances.

Uncontrolled Diabetes Mellitus

A 51-year-old woman had been known to have diabetes mellitus for ten years. She was checked by her physician every two months, and the disease remained in good control as a result of moderate dietary measures and the use of an oral hypoglycemic agent.

Three days before admission the patient developed a rather severe "cold." She had a mild fever, lost her appetite, and urinated more frequently, especially during the night. On the day of admission to the hospital she had begun to vomit and she felt short of breath.

Physical examination on admission revealed an acutely ill person whose mucous membranes and axillae were dry. She had periods of deep breathing (Kussmaul respiration). She was alert. Her temperature was 38.2°C, her pulse rate 100 per minute, her respiratory rate about 24 per minute, and her blood pressure 150/110 mm Hg. A urine specimen obtained on admission contained large amounts of sugar and ketones. The results of some other laboratory tests on admission are shown in Table 4-3.

Comment

The history is typical of uncontrolled diabetes mellitus, which is commonly set off by an intercurrent illness, such as an infection. The increased urination reflects an osmotic diuresis due to glucosuria, which was confirmed by large amounts of glucose in the urine. Uncontrolled diabetes mellitus often leads to primary metabolic acidosis, called ketoacidosis because it is due to an

Table 4-3 : Results of laboratory tests in a 51-year-old woman before and during treatment for diabetic ketoacidosis
The numbers in parentheses indicate estimated values.

Test	On admission	12 Hours after admission
Arterial blood:		
pH	7.15	7.43
Pco_2 (mm Hg)	13	23
$[H^+]$ (nmoles/L)	71 (72)	37 (37)
$[HCO_3^-]$ (mmoles/L)	4 (4)	15 (15)
Venous serum:		
$[Na^+]$ (mmoles/L)	130	142
$[Cl^-]$ (mmoles/L)	94	112
$[K^+]$ (mmoles/L)	3.7	3.5
Total CO_2 (mmoles/L)	5	16
Anion gap (meq/L)	31	14
Glucose (mg/100 ml)	762	156
Glucose (mmoles/L)	42.3	8.7

overproduction of ketoacids, mainly 3-hydroxybutyric and ace-toacetic acids (Fig. 4-1). This impression is supported by the arterial pH value, which is in the acidemic range (< 7.35). If the hyperventilation that the patient manifested had been the primary cause of the acid-base disturbance, the arterial pH should have been in the alkalemic range (> 7.45). Since it was not, the hyperventilation was probably a compensatory response to rather severe metabolic acidosis. If that is correct, the patient's Pco_2 should fall to a level predicted by the rule of thumb for uncomplicated metabolic acidosis (Table 3-3), and her arterial values should fall within the confidence bands in Figure 3-2. On the other hand, if the labored breathing reflected an added primary respiratory disturbance, one would expect the Pco_2 to fall by a greater amount than predicted by the rule of thumb, and the acid-base values to fall outside the bands. It can be seen that the fall of 27 mm Hg in Pco_2 from the normal value (40 − 13) is virtually identical to that of 26 mm Hg predicted by the rule of thumb (fall in $[HCO_3^-]$ from normal, 24 − 4 = 20; 20 × 1.3 = 26 mm Hg). Thus, it may be concluded that we are dealing with an uncomplicated metabolic acidosis with appropriate respiratory compensation. The increased anion gap, which was due to the accumulation of ketoanions in the plasma, further supports the diagnosis.

The low serum concentrations of Na^+ and Cl^- are often, but not invariably, seen in uncontrolled diabetes mellitus; they result partly from a shift of H_2O from the intracellular into the extracellular space and partly from the increased urinary excretion of NaCl that accompanies an osmotic diuresis.

Treatment

Severe diabetic ketoacidosis often leads to coma, as does severe hypoglycemia, which diabetic patients can develop from relatively too much insulin or other hypoglycemic agents. Therefore, when a diabetic patient is comatose it is important to ascertain the blood glucose concentration before giving the patient insulin. In the present case, insulin could be given immediately because the patient was alert; an intravenous infusion was started at once through the same needle from which the venous blood had been obtained. During the 12 hours after admission, a total of 3 liters of isotonic saline was given to replace urinary losses, and to those infusions were added a total of 250 units of crystalline insulin, 120 mmoles of KCl, and 200 mmoles of $NaHCO_3$. In addition, the patient was urged to drink liquids high in K^+, such as bouillon, tea, orange juice, and skim milk. Replacement of potassium is important because, as the hyperglycemia and acidosis are corrected, a shift of K^+ into cells can lead to dangerous hypokalemia.

The laboratory values after 12 hours of treatment are also shown in Table 4-3. Bicarbonate is distributed primarily in the extracellular space (ECF). Nevertheless, the apparent volume of distribution for administered HCO_3^- markedly exceeds the ECF because, as extracellular H^+ is buffered by the added HCO_3^-, H^+ moves out of cells. Movement of H^+ out of cells consumes a portion of the administered HCO_3^-, which has the same net effect as if HCO_3^- moved into the cells. Consequently, one calculates the amount of HCO_3^- to be given *as if* it were distributed into a volume greater than the ECF. This volume has been determined empirically to be approximately 50 percent of the body weight when serum bicarbonate concentration is normal or moderately decreased. The volume of distribution is even larger when plasma bicarbonate concentration is very low because intracellular buffers contribute a larger fraction to the total buffering of H^+ when acidosis is severe (for further explanation, see below, under Ingestion of Ethylene Glycol, Treatment).

In practice, one calculates how many mmoles per liter the $[HCO_3^-]$ is to be raised and multiplies this value by a volume equal to 50 percent of the body weight in kilograms. This computation is intended only as an approximate guide. In addition to entailing a rough estimate of the space of distribution, it assumes no further addition of acid to the body fluids, and no endogenous production of alkali. In most clinical settings, these assumptions are not completely valid, as exemplified by the patient under consideration. Administration of insulin to a patient with diabetic ketoacidosis not only stops the production of ketoacids but also permits regeneration of bicarbonate from metabolism of circulating ketoacid anions to CO_2 and water (Fig. 4-1). Assuming a body weight of 60 kg for the present patient, one would anticipate that the administration of 200 mmoles of $NaHCO_3$ would, at most, increase serum $[HCO_3^-]$ by 7 mmoles per liter (i.e., 200/(60 × 0.5)). In actuality, serum $[HCO_3^-]$ increased by 11 mmoles per liter (Table 4-3), indicating endogenous regeneration of bicarbonate. Further evidence for metabolism of ketoanions is afforded by the fall in the anion gap from 32 to 15 meq per liter, reflecting the reduction in the plasma concentration of these organic acid anions (Fig. 3-3). Because of these concurrent metabolic events, it is common practice to administer only enough bicarbonate for partial correction of the deficit (usually, to increase serum $[HCO_3^-]$ to 10 to 15 mmoles per liter).

An additional problem with administration of alkali is illustrated by this patient. As can be seen in Table 4-3, *partial* correction of $[HCO_3^-]$ *fully* restored the arterial pH to normal (in fact, even slightly beyond) because P_{CO_2} remained low. This phenomenon

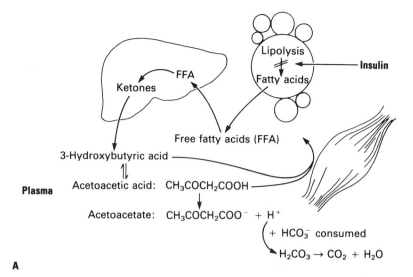

A

B

Fig. 4-1 : Dynamics of ketosis in diabetic ketoacidosis, as exemplified by acetoacetic acid: (A) Formation of ketoanions before insulin is given, and (B) metabolism of ketoanions after treatment with insulin has begun.

A. When insulin is not present, lypolysis in fat cells is increased. The resulting free fatty acids enter the liver, where they are preferentially converted to ketones, mainly 3-hydroxybutyric and acetoacetic acid. These ketoacids (here exemplified by acetoacetic acid) dissociate, forming ketoanions and H^+ at the pH of the body fluids. The H^+ reacts rapidly with bicarbonate, consuming this anion in the buffering process. Ketoanions rise in the plasma, not only because of increased production of ketoacids but also because their peripheral utilization, mainly by muscle, is diminished when insulin is lacking.

B. When insulin is given, lypolysis is slowed and entry of ketoacids into muscle (and other tissues) is accelerated. In the process HCO_3^- is regenerated.

Adapted from Witters, L. A., in May, H. L. (ed.), *Emergency Medicine.* New York: Wiley, 1984. Chap. 27.

occurs because the HCO_3^- in plasma does not equilibrate very quickly with the HCO_3^- in the cerebrospinal fluid (CSF). Consequently, the pH of CSF remains lower than the arterial pH, leading to continued ventilatory stimulation. Therefore, had more HCO_3^- been given to this patient, it is likely that an iatrogenic alkalosis might have complicated the picture (Eq. 3-6).

Because of the lag in reducing compensatory hyperventilation, the acid-base status 12 hours after admission indicated a mixed disturbance of metabolic acidosis and respiratory alkalosis. Accordingly, the arterial values now fall outside the confidence bands in Figure 3-2, as well as outside the bands for acute respiratory alkalosis (Fig. 3-1). The values do fit both the rule of thumb and the confidence bands for *chronic* respiratory alkalosis, which this patient did not have. This clinical example thus illustrates the important dictum that *rules of thumb and other aids must always be used within the context of the history and clinical setting.*

The increase in plasma chloride concentration seen after 12 hours reflects, in part, retention by the kidney of administered chloride. To the extent that ketoanions are lost in the urine during treatment, additional H^+ must be excreted as titratable acid or ammonium in order to regenerate the HCO_3^- that was originally consumed in buffering the ketoacids (Figs. 2-3, 2-6, and 4-1). The required increase in renal acid excretion often lags behind the loss of ketoanions from the plasma. This delay results in a sustained low plasma $[HCO_3^-]$, which is associated with a reciprocal increase in $[Cl^-]$ (see Chap. 3, under Meaning of Anion Gap, p. 79). The latter increment is due to a transient increase in the tubular reabsorption of Cl^-.

Once the acute imbalances of H_2O and electrolytes have been reversed, it is usually safer to curtail or discontinue vigorous intravenous therapy and to allow several days for the patient's renal and respiratory mechanisms to restore values fully to normal. During this period, however, careful observation of the patient, including frequent monitoring of the laboratory values, must be continued.

Lactic Acidosis

A 71-year-old woman was admitted to the hospital with fever and confusion. She had recently been discharged from another hospital after an inconclusive investigation for malaise. During that hospitalization, she was found to have abnormal liver function tests (enzymes), but the cause of those abnormalities was not discovered. After discharge she continued to feel weak and tired, and the day before the present admission she developed a fever and became disoriented.

Table 4-4 : Results of laboratory tests in a 71-year-old woman admitted with fever and confusion, who developed a lactic acidosis 24 hours after admission
Numbers in parentheses are estimated values.

Test	On admission	24 Hours after admission	6 Hours later	12 Hours later
Arterial blood:				
pH	7.50	7.14	7.16	7.18
P_{CO_2} (mm Hg)	30	18	20	22
$[H^+]$ (nmoles/L)	32 (32)	72 (74)	69 (70)	66 (67)
$[HCO_3^-]$ (mmoles/L)	23 (23)	6 (6)	7 (7)	8 (8)
P_{O_2} (mm Hg)		80		
Venous serum:				
$[Na^+]$ (mmoles/L)	138	140	142	144
$[K^+]$ (mmoles/L)	4.0	4.8	4.8	4.5
$[Cl^-]$ (mmoles/L)	105	105	105	106
Total CO_2 (mmoles/L)	25	6	8	9
Anion gap (meq/L)	8	29	29	29

On admission to the hospital her blood pressure was 100/60 mm Hg, temperature 39.5°C, and pulse rate 110 per minute and regular. She had a ventilatory rate of 18 per minute, and she was breathing somewhat deeply. She was confused and disoriented as to time and place. Her skin was warm and dry. She had no jaundice. Physical examination was otherwise unremarkable save for some diffuse lymphadenopathy.

Initial laboratory studies showed a hematocrit of 30%, white blood cell count of 5,800 per mm^3, and a serum creatinine concentration of 1.2 mg per 100 ml (106 μmoles per liter). Table 4-4 shows pertinent acid-base and electrolyte values on admission and during her hospitalization. Blood cultures were obtained, and antibiotics were started for presumed gram-negative sepsis. One day later she remained confused and appeared to be hyperventilating even more. Her temperature ranged from 38 to 39.5°C. Repeat laboratory measurements 24 hours after admission showed a dramatic change (Table 4-4). At that time serum ketones (a qualitative test for acetoacetate) were negative, serum glucose was 100 mg per 100 ml (5.6 mmoles per liter), and arterial P_{O_2} was 80 mm Hg.

Comment

At the time of admission this woman's history was suggestive of bacterial sepsis. She had a notable increase in the depth and rate of ventilation, and it was therefore suspected that her P_{CO_2} would be reduced. This change, which was substantiated by arterial blood gas analysis (Table 4-4), could have occurred either as a compensatory response to a metabolic acidosis or as a primary event reflecting a respiratory alkalosis. The fact that on admission the arterial pH was clearly in the alkalemic range showed conclusively that the patient had the latter disorder (see Chap. 7, under Gram-Negative Sepsis).

Although in this instance all the acid-base variables were supplied by the laboratory, it is advisable to estimate them as well, in order to exclude a laboratory or transcription error. In this instance the estimated $[H^+]$ was $0.8 \times 40 = 32$ nmoles per liter (Table 3-1), and the estimated $[HCO_3^-]$ (from Eq. 3-5) was $24 \times (30/32) = 23$ mmoles per liter. Both by the rule of thumb (Table 3-3) and by the confidence bands (Fig. 3-1), these data are consistent with respiratory alkalosis, a disorder that is commonly associated with sepsis (see Chap. 7 and Table 7-1). As stated in Chapter 3 (under Confidence Bands, p. 72), it is often difficult or impossible to separate acute from chronic respiratory alkalosis solely on the basis of laboratory data; since the present patient probably had hyperventilated for less than 24 hours, her disorder might still be classified as acute (see Chap. 1, last paragraph under Compensatory Responses).

Twenty-four hours after admission the acid-base status had changed dramatically. Arterial pH was now 7.14 (estimated $[H^+]$ $= 80 - 0.4 \times (80 - 64) = 74$ nmoles per liter), and the $[HCO_3^-]$ was 6 mmoles per liter (estimated $[HCO_3^-] = 24 \times (18/74) = 6$ mmoles per liter). The patient now had a severe acidosis, and since the P_{CO_2} was low, it must have been a metabolic acidosis. Using the rule of thumb for this disorder (Table 3-3), one would predict a *change* in P_{CO_2} (ΔP_{CO_2}) of $1.3 \times (24 - 6)$, or 23 mm Hg, and hence a new P_{CO_2} of $40 - 23$, or 17 mm Hg. The P_{CO_2} in this patient 24 hours after admission was thus consistent with the compensatory response seen in uncomplicated metabolic acidosis (see also Fig. 3-2). Note that in the above calculation we have used normal values for $[HCO_3^-]$ and P_{CO_2} in computing the changes (Δ) for these variables ($24 - 6$ and $40 - 23$, respectively). The rules of thumb apply to aberrations from *normal* in the acid-base disturbance under consideration, not to the changes that occur when one moves from one type of disturbance to another.

In the absence either of ingestion of a large amount of acid or of a tremendous loss of HCO_3^- in diarrheal stools (neither of which the patient manifested), the precipitous fall in $[HCO_3^-]$ that occurred during the first 24 hours of hospitalization (Table 4-4) must be ascribed to a dramatic increase in the endogenous production of fixed acids. Under conditions of normal fixed acid production, even total cessation of the renal excretion of acid would lead to a drop in plasma $[HCO_3^-]$ of only 2 to 3 mmoles per liter in 24 hours. This deduction follows from the fact that roughly 50% of the daily load of fixed acid, say 50 mmoles (p. 1), would be buffered by extracellular HCO_3^- (Fig. 1-6), which in a healthy 70-kg person totals approximately 336 mmoles (24 mmoles/L $\times$ 14 L of extracellular fluid). Consistent with the conclusion that production of endogenous acid must have increased in this patient is the finding that her anion gap rose from 8 to 29 meq per liter (Table 4-4). This rise indicated that some endogenous acid was titrated by body buffers, leaving behind a neutral salt whose anion moiety is not routinely measured (as in the prototypical Eq. 1-5). The question then is, which acid and which anion? The negative test for plasma ketones, as well as the normal blood glucose concentration, exclude the ketoanions, 3-hydroxybutyrate and acetoacetate. The next most common candidate is lactate, arising from the buffering of lactic acid, and this suspicion was confirmed by elevation of the blood lactate concentration from the normal value of approximately 1 mmole per liter (Table 1-1) to 15 mmoles per liter. In fact, this elevation could account for the entire increase in the anion gap in this patient.

The causes of lactic acidosis are listed in Table 4-5. The present patient was not hypoxemic, as witness her normal Po_2, nor was she in shock from her infection. At the time of admission she manifested none of the conditions listed in Table 4-5 as leading to increased production of lactic acid. However, her liver function tests had been abnormal during the previous hospitalization, and subsequently it was found that she had a disseminated lymphoma with extensive involvement of the liver. Thus, the most likely explanation for this patient's lactic acidosis is that she had a combination of increased lactic acid production by the lymphoma and some impairment of lactate removal from the circulation, since the liver is a major site for the metabolism of

Table 4-5 : Causes of lactic acidosis

Increased Production	Decreased Removal
Severe hypoxemia	Severe hepatic failure
($Po_2 < 30$ mm Hg)	
Shock	
Generalized convulsions	
Severe voluntary exercise	
Disseminated neoplasms	
Leukemias, lymphomas	
Drugs	
Phenformin	
Ethanol	
Associated illnesses	
Diabetes mellitus	
Sepsis	
Idiopathic	

lactate. The latter process — impairment of lactate removal — would aggravate the acidosis, since metabolism of lactate to CO_2 and water would ordinarily regenerate HCO_3^- (for an analogous situation, see Fig. 4-1B).

Treatment

As soon as the presence of severe metabolic acidosis ([HCO_3^-] < 10 mmoles per liter) was documented, the patient was given 200 mmoles of $NaHCO_3$ intravenously over the next several hours. This step was accomplished by the addition of 2 ampules of concentrated $NaHCO_3$ (Fig. 1-8C) to each of two liters of 5% dextrose in half-strength saline (Fig. 1-8A). The patient weighed 55 kg and, if one assumes a volume of distribution for HCO_3^- equal to 50% of the body weight, this treatment should have increased plasma [HCO_3^-] by approximately 7 mmoles per liter (200 mmoles/27.5 liters). Six hours later, however, the plasma [HCO_3^-] had risen minimally if at all, and there was little change in the other values on blood gas analysis (Table 4-4). Even after an additional 300 mmoles of $NaHCO_3$ had been given intravenously, the plasma [HCO_3^-] had increased to only 8 mmoles per liter (Table 4-4, 12 hours later). This failure of plasma [HCO_3^-] to rise as predicted reflects the continued production of large amounts of lactic acid in excess of its removal by metabolism. Thus, as the net production of lactic acid continues to add H^+ to the body fluids, the administered alkali will continue to be consumed in buffering the added acid (Eq. 1-5). Reflecting this process, blood lactate concentration, and hence the anion gap, will remain high. In the present patient the plasma [HCO_3^-] ranged between 8 and 10 mmoles per liter over the next several days, regardless of the amount of HCO_3^- that was administered, indicating unabated production of lactic acid. The patient died several days after admission, despite continued intensive therapy.

This patient reflects a common experience with lactic acidosis. In contrast to ketoacidosis, where appropriate therapy can promptly correct overproduction of fixed acid, one often cannot stop the increased production in lactic acidosis because the fundamental disease process is not readily reversible. In the absence of specific therapy, repletion with alkali is the only form of treatment presently available, tiding the patient over until production of acid spontaneously diminishes. Unfortunately, recent experimental studies have provided evidence that, for unknown reasons, $NaHCO_3$ may itself stimulate lactic acid production. Studies are under way to explore alternative approaches to therapy.

Ingestion of Ethylene Glycol

A 56-year-old man was found lying in an alley, unresponsive and breathing heavily. A half-empty can of antifreeze was found nearby. The man was known in the neighborhood to be a heavy

4. Metabolic Acidosis: Clinical Examples : 99

Table 4-6 : **Results of laboratory tests in a 51-year-old man suspected of ingesting ethylene glycol**
The numbers in parentheses indicate estimated values.

Test	On admission	4 Hours later
Arterial blood:		
pH	6.90	7.05
P_{CO_2} (mm Hg)	11	15
[H^+] (nmoles/L)	126 (125)	89 (90)
[HCO_3^-] (mmoles/L)	2 (2)	4 (4)
Venous serum:		
[Na^+] (mmoles/L)	142	145
[K^+] (mmoles/L)	5.0	4.5
[Cl^-] (mmoles/L)	103	103
Total CO_2 (mmoles/L)	3	5
Anion gap (meq/L)	36	37

consumer of alcohol, and he was frequently seen panhandling for money to buy liquor. He was brought to the emergency room, where he was found to be comatose. He showed a normal pupillary reaction to light, and he withdrew in response to painful stimuli. There was obvious Kussmaul breathing, with a respiratory rate of 36 per minute. Blood pressure was 120/60 mm Hg, and pulse rate was 70 per minute and regular. Laboratory values on admission and after treatment are shown in Table 4-6.

Comment　　In our discussion we will proceed from diagnosis to treatment, even though in this instance treatment was begun immediately because of the life-threatening reduction in systemic pH (see Chap. 1, first paragraph). Because of the clinical history (an alcoholic patient in coma, with Kussmaul respiration; the half-empty can of antifreeze), the physician in the emergency room surmised that she was most probably dealing with severe metabolic acidosis due to ingestion of ethylene glycol. She therefore started an intravenous infusion of $NaHCO_3$ at once and then asked for a blood gas analysis. As soon as the laboratory technician measured an arterial pH of 6.90, he phoned this information to the physician, who then, by quick estimation, could gauge the severity of the patient's acid-base disturbance. To estimate the [H^+] at a pH of 6.90, the value for 7.00 (namely, 100 nmoles per liter) is divided by 0.8, yielding 125 nmoles per liter (Table 3-1). With severe Kussmaul respiration, the rate of alveolar ventilation is likely to be tripled or quadrupled, which results in a P_{CO_2} of 13 to 10 mm Hg. Thus, the [HCO_3^-] could be estimated as, say, 24 × 12/125, or 2 mmoles per liter. Even if the P_{CO_2} were as high as 20 mm Hg, the [HCO_3^-] would be only 4 mmoles per liter (24 × 20/125). This estimate confirmed the physician's suspicion that she was dealing with an emergency. The measured values on admission (Table 4-6), which were reported later, confirmed the correctness of her estimate.

The combination of a low $[HCO_3^-]$ and pH indicates the presence of a metabolic acidosis. The compensatory ventilatory response, which reduces the PCO_2 to 11 mm Hg, is of a degree to be expected in a severe but otherwise uncomplicated metabolic acidosis: The patient's $[HCO_3^-]$ decreased by $24 - 2$, or 22 mmoles per liter, and the predicted change in PCO_2 would therefore be 1.3×22, or 29 mm Hg (Table 3-3), yielding a new PCO_2 of $40 - 29$, or 11 mm Hg. Although the prediction is fulfilled in this patient, the rules of thumb must be applied with caution in very severe metabolic acidosis ($[HCO_3^-] < 6$ mmoles per liter). The observations on which the rules are based were obtained from subjects whose $[HCO_3^-]$ was approximately 10 mmoles per liter or higher; note that the confidence bands (Fig. 3-2) do not extend lower than a HCO_3^- concentration of 5 mmoles per liter. In fact, a PCO_2 of 11 mm Hg may represent a near maximal ventilatory effort, possibly — and this is not proven — because this is about the level where any further ventilatory effort, and hence further production of CO_2, will prevent any additional net increase in CO_2 excretion. Note also that the confidence bands for primary respiratory disturbances (Fig. 3-1) do not extend below a PCO_2 of 15 mm Hg.

The anion gap in this patient was 36 meq per liter (Table 4-6). Even though this finding further strengthened the physician's suspicion that the patient had ingested ethylene glycol, she appropriately obtained additional studies to exclude other, more common causes of metabolic acidosis with an increased anion gap (Table 4-1). Serum ketones were normal, as was serum glucose concentration (90 mg per 100 ml, or 5 mmoles per liter). The serum concentration of lactate was only 2 mmoles per liter, which could not possibly account for the elevation of the anion gap by approximately 24 meq per liter. Serum creatinine concentration was also normal. The measured serum osmolality was 374 mosmoles per kg H_2O, i.e., some 80 mosmoles per kg H_2O higher than the value predicted from the concentrations of the major solutes (serum $[Na^+] \times 2$, plus glucose and urea in mmoles per liter). The excess of measured over predicted osmolality indicated the presence in the serum of an unmeasured solute of low molecular weight. Eventually, ethylene glycol (M.W. = 62 daltons) was identified in the serum, although this analysis was not available at the time of admission.

Ethylene glycol ($C_2H_6O_2$) itself is neither toxic nor acid. In the body, however, it is metabolized to a series of toxic acids, including glycolic acid, glyoxylic acid, oxalic acid, and formic acid. When these acids are buffered, the resulting neutral salts of these acids (Eq. 1-5) contribute to the pool of "unmeasured anions" and thereby raise the anion gap.

Treatment

The initial aim of therapy was to raise the plasma $[HCO_3^-]$ quickly to above 10 mmoles per liter. Therefore, 400 mmoles of $NaHCO_3$ (8 ampules, Fig. 1-8C) in 2 liters of 5% dextrose in water was infused intravenously during the first hour after admission. With the patient weighing 66 kg, and the estimated volume of distribution of administered HCO_3^- equal to 50% of the body weight, the predicted rise in plasma $[HCO_3^-]$ would be 400 mmoles/33, or approximately 12 mmoles per liter. As Table 4-6 shows, however, this goal was not achieved even after more $NaHCO_3$ was infused. There are probably two reasons for this failure. One is continued production of acids from the metabolism of ethylene glycol; this fact is reflected in the persistently high anion gap. The other reason is that when, as in this patient, the plasma $[HCO_3^-]$ is very low, intracellular, nonbicarbonate buffers are titrated extensively by the accumulating H^+ ions (Fig. 1-6). When HCO_3^- is administered in this setting, release of H^+ from these intracellular buffers consumes a large fraction of the added alkali. As a result, the apparent volume of distribution for administered HCO_3^- is much larger than 50% of body weight (see Uncontrolled Diabetes Mellitus, Treatment, this chapter). When the serum $[HCO_3^-]$ is as low as 2 mmoles per liter, the volume of distribution approaches 90% of body weight.

Despite the small increment in plasma $[HCO_3^-]$, the treatment may well have saved the patient's life, for an arterial pH of 7.05, although still dangerous, is much more compatible with survival than a pH of 6.90. The relatively large effect of a small change in plasma $[HCO_3^-]$ on the $[H^+]$ is not surprising if one considers the relationships in the Henderson equation, namely, $[H^+] = 24 \times Pco_2/[HCO_3^-]$ (Eq. 3-4). It follows from this equation that when the denominator is very small initially, even a seemingly trivial increase may greatly alter the ratio, $Pco_2/[HCO_3^-]$. Thus, in the present patient, where the denominator doubled (albeit from only 2 to 4 mmoles per liter), the $[H^+]$ would have fallen to half its initial value if the Pco_2 had not increased simultaneously. Probably the Pco_2 rose slightly because with an increased pH there was somewhat less respiratory drive. Note that the values at 4 hours in Table 4-6 illustrate the quick check on the accuracy and consistency of acid-base values that we described in conjunction with Equation 3-4. We said there (p. 66) that when the Pco_2 is four times the value for $[HCO_3^-]$, the pH should be approximately 7.00.

Several other therapeutic measures were instituted as soon as the intravenous infusion of $NaHCO_3$ had been started: (1) The patient was intubated, in order to assure free ventilation and to prevent aspiration of gastric contents; (2) a nasogastric tube was inserted into the stomach, which was rinsed several times in an effort to remove any unabsorbed ethylene glycol; (3) mannitol was added to the intravenous infusion (Fig. 1-8C) to induce an osmotic diuresis, which enhances the urinary excretion of ethylene glycol by retarding its renal tubular reabsorption; (4) ethyl alcohol was administered by intravenous infusion in an amount sufficient to raise the serum alcohol level to over 100 mg per 100 ml. The enzyme alcohol dehydrogenase metabolizes both ethyl alcohol (C_2H_5OH) and ethylene glycol ($C_2H_6O_2$), but, since the enzyme has a higher affinity for the first compound than for the second, ethyl alcohol retards the metabolism of ethylene glycol and, hence, the production of the toxic acids; finally, (5) hemodialysis was carried out, both to remove ethylene glycol and to permit the addition of further amounts of HCO_3^- to the body without inducing hypertonicity of the plasma.

As the result of this aggressive therapy, the patient's acid-base status returned to normal during the three days following admission. He was then transferred to a rehabilitation unit for alcoholics.

Diarrhea

A 50-year-old businessman developed very severe diarrhea while visiting a foreign capital in the tropics. At first he and his wife tried to manage the disorder by themselves, using an anti-diarrheal drug containing paregoric that they had brought from home, and having him take clear liquids. However, he continued to pass watery stools at the rate of some 12 to 20 times during a 24-hour period. He became progressively weaker, and on the third day his wife took him to the emergency room of a medical school-affiliated hospital.

On examination, his eyes appeared sunken, his mucous membranes were dry, and he had poor skin turgor. His blood pressure was 90/60 mm Hg while he lay supine, and it fell to unmeasurably low levels when he sat up. His pulse rate was 120 per minute and regular, and his respirations were notably deep at approximately 24 per minute. He was admitted to the hospital.

Results of laboratory tests are shown in Table 4-7. Although blood gas analysis was not routinely available at this hospital, it was possible to determine the pH on arterialized venous blood (i.e., on blood from a superficial vein on the back of the hand after immersing the hand in hot water).

Comment

On the basis of the history, one would predict that the patient has a metabolic acidosis caused by loss of HCO_3^- in the stool (Table 1-3). This prediction was borne out by the acid pH of the arterialized venous blood (Table 4-7), as well as by the normal

Table 4-7 : Results of laboratory tests
in a 50-year-old man with severe diarrhea
The numbers in parentheses indicate estimated values.

Test	On admission	12 Hours later
Arterialized venous blood:		
pH	7.25	—
P_{CO_2} (mm Hg)	(24)	—
$[H^+]$ (nmoles/L)	(57)	—
$[HCO_3^-]$ (mmoles/L)	(10)	—
Venous serum:		
$[Na^+]$ (mmoles/L)	132	139
$[K^+]$ (mmoles/L)	2.3	3.0
$[Cl^-]$ (mmoles/L)	111	112
Total CO_2 (mmoles/L)	12	17
Anion gap (meq/L)	9	10

anion gap (Table 4-1). Although the other acid-base values were not furnished by the laboratory, it was possible to estimate them in the following manner: The $[H^+]$ could be estimated as 50 + $[0.5 \times (64 - 50)]$, or 57 nmoles per liter. The arterialized $[HCO_3^-]$ was estimated simply as being 2 mmoles per liter lower than the total CO_2 measured on venous blood (see p. 62), i.e., 10 mmoles per liter. Then the P_{CO_2} could be estimated from Equation 3-4 as $P_{CO_2} = [H^+] \times [HCO_3^-]/24 = 57 \times 10/24$, or 24 mm Hg.

By the rules of thumb (Table 3-3) one would have predicted a compensatory fall in P_{CO_2} of $(24 - 10) \times 1.3$, or 18 mm Hg; the new P_{CO_2} should therefore be $40 - 18$, or 22 mm Hg. The estimated value of 24 mm Hg (Table 4-7) thus was consistent with the presence of uncomplicated metabolic acidosis in this patient (see also Fig. 3-2). The other causes of metabolic acidosis with a normal anion gap (Table 4-1) were so unlikely that further investigation for them was not warranted unless appropriate therapy (see below) failed to correct the acid-base disturbance.

The patient's postural hypotension reflected rather severe contraction of the body fluid volumes, especially of the extracellular space. Severe diarrhea leads to disproportionate losses not only of HCO_3^- but also of K^+ (Table 1-3). This fact accounts for the hypokalemia on admission (Table 4-7). The disproportionate loss of HCO_3^- from the extracellular fluid and the associated volume contraction produced the observed changes in the composition of serum anions, i.e., a reduction in serum total CO_2 and a reciprocal increase in $[Cl^-]$ (see Chap. 3, under Meaning of Anion Gap, p. 79). These reciprocal changes are induced by the selective losses of ions and by volume contraction, but they are sustained by consequent changes in renal reabsorption, due to volume-stimulated avidity for Na^+. Because less filtered HCO_3^- is available to accompany the reabsorbed Na^+, reabsorption of

Cl^- increases, thereby maintaining the hyperchloremia. It is important to emphasize, however, that despite the presence of hyperchloremia, net losses of Cl^-, as well as of HCO_3^-, occur via the stool as a result of the diarrhea. Therefore, treatment should include administration not only of HCO_3^- and K^+, but also of NaCl — all, of course, with water, since volume needs to be replenished.

Treatment

Replacement of the losses described above is accomplished by giving the patient 3 or 4 liters of normal saline or half-strength saline (Fig. 1-8A), to which KCl and $NaHCO_3$ are added (Fig. 1-8C). There are no hard-and-fast quantitative rules for correcting extracellular fluid volume. In younger patients who have no evidence of cardiac or renal impairment, the adequacy of fluid replacement can be gauged by monitoring blood pressure, pulse rate, skin turgor, and output of urine. In older patients, however, especially if they have a history of cardiac or renal insufficiency, it is safer to monitor the ability of the left ventricle to handle the replenished extracellular volume. This goal can be accomplished by passing a small pressure-sensing catheter (via a major vein) into the pulmonary artery and periodically wedging it into a pulmonary capillary (a so-called Swan-Ganz line). Inasmuch as pulmonary capillary pressure rises when the left ventricle fails, overload with administered fluid will be detected by an elevation of the wedge pressure.

The patient under consideration was given 200 mmoles of $NaHCO_3$ during the first six hours after admission. He weighed 68 kg, and it would therefore be expected that his plasma $[HCO_3^-]$ would rise by approximately 6 mmoles per liter, i.e., $200/(0.5 \times 68)$; this prediction agrees well with the measured rise of 5 mmoles per liter (Table 4-7). Apparently, then, no further losses of HCO_3^- were occurring, nor was there an unsuspected excessive production of acid. In contrast to diabetic ketoacidosis and lactic acidosis, where metabolism of the applicable anions can regenerate HCO_3^- rapidly, in diarrhea the HCO_3^- is lost from the body. Replenishment of the body stores of HCO_3^- can be accomplished only by an increase in the renal excretion of acid, a response that takes several days. Hence, in diarrhea-induced metabolic acidosis $NaHCO_3$ is necessary in order to avoid a severe and prolonged acidosis, even if the diarrhea abates.

In this patient the diarrhea stopped shortly after admission. Beginning with the second hospital day, he was cautiously restarted on clear liquids by mouth, and the rate of intravenous fluid administration was decreased and then discontinued. He was then rapidly advanced to a normal diet and was discharged on the fourth hospital day.

Renal Tubular Acidosis

A 23-year-old woman consulted a physician because during the previous year she had experienced a gradual loss of energy, as well as weakness and fatigue. She had been apparently well up to a year earlier, and she had not been seen by a physician for years. There was no history of anorexia, weight loss, nausea, vomiting, or diarrhea, and she was not taking any medication.

On physical examination she had a normal blood pressure, pulse, and respiratory rate. The only abnormal finding was symmetrically hypoactive deep tendon reflexes. Her serum creatinine concentration was 1.0 mg per 100 ml (88 μmoles per liter) and her hematocrit was 40%. Other laboratory values are shown in Table 4-8.

Comment

The combination of a low arterial pH and a low plasma $[HCO_3{}^-]$ indicates the presence of a metabolic acidosis. In contrast to the other patients discussed in this chapter, this patient's acid-base disorder could not have been suspected from the clinical history and the physical examination. Although fatigue and weakness occur with renal tubular acidosis, these symptoms are nonspecific and the disorder is quite rare. Screening laboratory tests, such as the plasma electrolytes shown in Table 4-8, were required to raise suspicions of the correct diagnosis and to reveal the cause of the patient's symptoms, which were due both to the presence of metabolic acidosis and to the associated hypokalemia. As will be discussed below, hypokalemia is a characteristic and diagnostic feature of the type of metabolic acidosis that this patient has (i.e., distal renal tubular acidosis).

In Chapter 3, under Estimation of $[HCO_3{}^-]$ (p. 66), we noted that whenever the P_{CO_2} is twice the $[HCO_3{}^-]$, the pH will be approximately 7.30; the values listed in Table 4-8 thus illustrate how this

Table 4-8 : Results of laboratory tests in a 23-year-old woman with renal tubular acidosis
The numbers in parentheses indicate estimated values.

Test	On first evaluation	After treatment
Arterial blood:		
pH	7.30	—
P_{CO_2} (mm Hg)	28	—
$[H^+]$ (nmoles/L)	50 (50)	—
$[HCO_3{}^-]$ (mmoles/L)	14 (13)	—
Venous serum:		
$[Na^+]$ (mmoles/L)	142	140
$[K^+]$ (mmoles/L)	2.8	3.0
$[Cl^-]$ (mmoles/L)	115	100
Total CO_2 (mmoles/L)	15	24
Anion gap (meq/L)	12	16
Urine:		
pH	6.4	

rule can serve as a quick check on the consistency of the laboratory data. With a fall in plasma $[HCO_3^-]$ of $(24 - 14)$, or 10 mmoles per liter, the rule of thumb (Table 3-3) predicts a compensatory decrease in P_{CO_2} of 1.3×10, or 13 mm Hg, yielding a new P_{CO_2} of $(40 - 13)$, or 27 mm Hg. The values on first evaluation (Table 4-8) are thus consistent with an uncomplicated primary metabolic acidosis. In this instance, the anion gap is normal because serum $[Cl^-]$ increased in a reciprocal fashion as the serum $[HCO_3^-]$ declined (see Chap. 3, under Meaning of Anion Gap, p. 79). With a normal anion gap, we then turn to the second part of Table 4-1 to find a cause for this patient's acid-base disorder. Since the patient has no history of diarrhea, intake of acidifying drugs, or ureterosigmoidostomy, renal tubular acidosis (RTA) emerges as the most likely diagnosis.

Renal tubular acidosis is a discrete tubular disorder (often without diminution of overall renal function) characterized by an impairment in the renal secretory capacity for hydrogen ion. In some instances the disease is inherited or appears as an idiopathic disorder, and in other instances it occurs in association with a variety of systemic disorders or as part of drug-induced or toxin-induced nephropathy. In this patient it is apparently either inherited or idiopathic.

Although many variants of RTA have been described, the disorder can be subdivided into two major classes, proximal and distal RTA. As discussed in Chapter 2, two tasks are required by the kidney to maintain acid-base balance: (1) reabsorption of all the filtered HCO_3^-, which occurs primarily in the proximal tubules, and (2) titration of urinary buffers and formation of ammonium (Figs. 2-3 and 2-6), processes that require the secretion of H^+ against a steep electrochemical gradient in the distal tubules and collecting ducts (see Chap. 2, Factors Affecting the Rate of T.A. Excretion, point 1). It is important to realize that these two functions are linked. In order for excretion of H^+ to proceed normally, virtually all the filtered HCO_3^- must be reabsorbed prior to the distal tubules, so that the H^+ that is secreted by the distal tubules and collecting ducts can serve to reduce the pH of tubular fluid. That reduction, in turn, is required for titration of Na_2HPO_4 (Figs. 2-3 and 2-5) and for trapping of NH_4^+ (Figs. 2-6 and 2-7), the two mechanisms by which H^+ is excreted. Recall from the Henderson-Hasselbalch equation (see Answer to Problem 2-3) that urinary $[HCO_3^-]$ must decrease to less than 1 mmole per liter if urinary pH is to fall below 6.0.

In proximal RTA, reabsorption of the filtered HCO_3^- is impaired, but distal acidification and, hence, the ability to excrete H^+ can be normal, provided that distal delivery of HCO_3^- is low (discussed next). In this disorder HCO_3^- escapes reabsorption in the proximal tubules (presumably due to a defect in hydrogen ion secretion in this segment of the nephron) and is carried into the distal tubule and collecting duct. Although H^+ secretion can proceed against a steep gradient in the distal nephron, the capacity to increase H^+ secretion in response to an increase in the delivery of HCO_3^- is quite limited. As a result, much of the excess HCO_3^- leaving the proximal tubule escapes into the urine (Fig. 4-2D). This process of wasting HCO_3^- continues until serum $[HCO_3^-]$ and hence the filtered load of HCO_3^- (GFR $\times$ $P_{HCO_3^-}$) fall sufficiently so that virtually all the filtered HCO_3^- can again be recaptured prior to the distal nephron (Fig. 4-2B). When this level of lowered serum $[HCO_3^-]$ has been reached, urinary acidification and hence H^+ excretion return to normal. Therefore, in the steady state of proximal RTA, serum $[HCO_3^-]$ remains low but the daily burden of fixed acid (see beginning of Chap. 1) can be excreted in the urine. There is, in this new state, no positive balance for acid, although systemic acidosis resulting from retention of H^+ prior to the attainment of the steady state persists; in the new steady state, urinary pH is low (<5.5). If, however, serum $[HCO_3^-]$ is raised through administration of HCO_3^-, excess HCO_3^- is again excreted in the urine and the urinary pH rises to abnormal levels. These dynamic aspects have been summarized in Table 4-9 and Figure 4-2.

By contrast, in distal RTA the reabsorption of HCO_3^- in the proximal tubules is normal or nearly normal, but the secretion of H^+ in the distal tubules and collecting ducts is impaired. As a result, despite virtually complete reabsorption of filtered HCO_3^-, excre-

Table 4-9 : Differences between the proximal
and distal forms of renal tubular acidosis (RTA)

Aspect	Proximal RTA	Distal RTA
Metabolic acidosis	Present	Present
Urinary HCO_3^- excretion:		
During acidosis	Nil	Slight ↑
With corrected $[HCO_3^-]$	Marked ↑	Slight ↑
Urinary pH:		
During acidosis	< 5.5	> 5.5
With corrected $[HCO_3^-]$	> 5.5	> 5.5
Urinary excretion of titratable acid (T.A.) and NH_4^+:		
During acidosis	Normal	↓ ↓
With corrected $[HCO_3^-]$	↓ ↓	↓ ↓
Serum $[K^+]$:		
During acidosis	Normal	↓
With corrected $[HCO_3^-]$	↓	Improved, but still ↓

Proximal Renal Acidification Defect

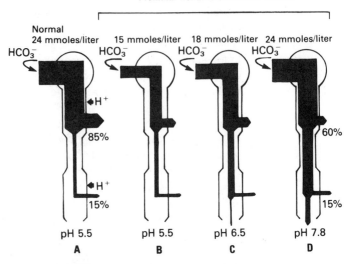

Fig. 4-2 : Dynamics whereby deficient reabsorption of filtered HCO_3^- in the proximal tubules, as in proximal RTA, can result in minimal urinary pH during steady-state acidosis (plasma HCO_3^- concentration = 15 mmoles per liter), but abnormally high urinary pH at lesser degrees of acidosis or when serum $[HCO_3^-]$ is restored to normal by administration of HCO_3^-. From Morris, R. C., Jr., *Calif. Med.* 108:225, 1968.

tion of the daily load of acid is incomplete and there is a positive balance for H^+, which leads to metabolic acidosis. The hallmark of impaired H^+ secretion in the distal nephron is a defect in urinary acidification (i.e., in the reduction of urinary pH); consequently, despite the presence of moderate to severe systemic acidosis, the urinary pH remains inappropriately high (Table 4-9).

In this patient the diagnosis of distal RTA was easily established by measuring the urinary pH. In normal individuals, a reduction in plasma $[HCO_3^-]$ below 18 mmoles per liter by metabolic acidosis causes urinary pH to fall to less than 5.5. The same is true of patients with proximal RTA when they are in steady-state metabolic acidosis. In this patient the urinary pH was 6.4 even though the plasma $[HCO_3^-]$ was 14 mmoles per liter (Table 4-8), indicating that the patient has distal RTA. Note from Table 4-9 that the three major differences between proximal RTA and distal RTA (namely, urinary excretion of HCO_3^-, urinary pH, and urinary excretion of titratable acid and NH_4^+) all depend on the presence of acidosis (see Problem 2-3 and the answer thereto).

As noted above, patients with proximal RTA can excrete their daily load of acid normally *once they have reached a steady state of acidosis,* whereas patients with distal RTA are never able to excrete acid adequately. This distinction has important ramifications for the complications of these disorders. In patients with distal RTA, the retained acid is buffered continuously by $CaCO_3$ in bone (yielding Ca^{2+} and HCO_3^-), and this process leads to loss of calcium from bone (osteomalacia) and to nephrocalcinosis and renal damage. In fact, an abdominal roentgenogram in the present patient revealed deposition of calcium in the kidney. In contrast, patients with proximal RTA do not develop bone disease and nephrocalcinosis unless other tubular defects are present.

Another feature that separates distal from proximal RTA is the *invariable* presence of hypokalemia in the former disorder (Table 4-9). The hypokalemia is due to renal losses of K^+. The precise mechanism for renal K^+ wasting in distal RTA is unknown, but two factors are postulated as main contributors to this phenomenon. First, these patients often manifest some volume contraction and stimulation of the renin-angiotensin-aldosterone system; aldosterone augments K^+ secretion. Second, an alkaline distal tubular fluid and continued delivery of sodium to the distal tubule and collecting duct are known to facilitate K^+ secretion. Both are present in distal RTA. In proximal RTA, on the other hand, urinary wasting of HCO_3^- occurs only when the patient is treated with $NaHCO_3$, a maneuver that alkalinizes the distal tubular fluid (Table 4-9). Therefore, only in this circumstance is hypokalemia seen in proximal RTA (Table 4-9).

Treatment

It follows from the above discussion that treatment with alkali may not be required in proximal RTA. In fact, most adult patients are not treated, although children, even with mild acidosis, are treated in order to prevent lag of growth and development. But such treatment is essential in distal RTA in order to maintain acid-base balance and hence to prevent complications such as osteomalacia, nephrocalcinosis, possible renal failure, and K^+ depletion. One need give only sufficient HCO_3^- to neutralize that portion of endogenously produced acids which cannot be excreted. In the present patient, after repletion of body HCO_3^- stores, a daily intake of 40 mmoles of HCO_3^- was enough to maintain her plasma $[HCO_3^-]$ above 20 mmoles per liter and to maintain normal acid-base balance. The HCO_3^- was given partly as the potassium salt (to correct the K^+ deficiency) and partly as the sodium salt. The patient's symptoms cleared, and she regained a feeling of well-being. Her disease was identified early enough to avoid bone disease and other complications. However, unless some new treatment is discovered, she will have to remain on the medication for the rest of her life.

110

Selected References

General

Adrogué, H. J., Brensilver, J., Cohen, J. J., and Madias, N. E. Influence of steady-state alterations in acid-base equilibrium on the fate of administered bicarbonate in the dog. *J. Clin. Invest.* 71:867, 1983.

Bushinsky, D. A., Coe, F. L., Katzenberg, C., Szidon, J. P., and Parks, J. H. Arterial P_{CO_2} in chronic metabolic acidosis. *Kidney Int.* 22:311, 1982.

Cohen, J. J., and Kassirer, J. P. *Acid-Base.* Boston: Little, Brown, 1982. Chap. 8.

Gabow, P. A. Disorders associated with an altered anion gap. *Kidney Int.* 27:472, 1985.

Garella, S., Dana, C. L., and Chazan, J. A. Severity of metabolic acidosis as a determinant of bicarbonate requirements. *N. Engl. J. Med.* 289:121, 1973.

Rose, B. D. *Clinical Physiology of Acid-Base and Electrolyte Disorders* (2nd ed.). New York: McGraw-Hill, 1984. Chap. 18.

Chronic Renal Failure

Buerkert, J. D., Martin, D., Trigg, D., and Simon, E. Effect of reduced renal mass on ammonium handling and net acid formation by the superficial and juxtamedullary nephron of the rat. *J. Clin. Invest.* 71:1661, 1983.

Fine, L. G., Trizna, W., Bourgoignie, J. J., and Bricker, N. S. Functional profile of the isolated uremic nephron. Role of compensatory hypertrophy in the control of fluid reabsorption by the proximal straight tubule. *J. Clin. Invest.* 61:1508, 1978.

Goodman, A. D., Lemann, J., Jr., Lennon, E. J., and Relman, A. S. Production, excretion, and net balance of fixed acid in patients with renal acidosis. *J. Clin. Invest.* 44:495, 1965.

Litzow, J. R., Lemann, J., Jr., and Lennon, E. J. The effect of treatment of acidosis on calcium balance in patients with chronic azotemic renal failure. *J. Clin. Invest.* 46:280, 1967.

Lubowitz, H., Purkerson, M. L., Rolf, D. B., Weisser, F., and Bricker, N. S. Effect of nephron loss on proximal tubular bicarbonate reabsorption in the rat. *Am. J. Physiol.* 220:457, 1971.

Maddox, D. A., Horn, J. F., Famiano, F. C., and Gennari, F. J. Load dependence of proximal tubular fluid and bicarbonate reabsorption in the remnant kidney of the Munich-Wistar rat. *J. Clin. Invest.* 77:1639, 1986.

Pellegrino, E. D., and Biltz, R. M. The composition of human bone in uremia. Observations on the reservoir functions of bone and demonstration of a labile fraction of bone carbonate. *Medicine* (Baltimore) 44:397, 1965.

Schwartz, W. B., Hall, P. W., III, Hays, R. M., and Relman, A. S. On the mechanism of acidosis in chronic renal disease. *J. Clin. Invest.* 38:39, 1959.

Trizna, W., Yanagawa, N., Bar-Khayim, Y., Houston, B., and Fine, L. G. Functional profile of the isolated uremic nephron. Evidence of proximal tubular "memory" in experimental renal disease. *J. Clin. Invest.* 68:760, 1981.

Van Slyke, D. D., Linder, G. C., Hiller, A., Leiter, L., and McIntosh, J. F. The excretion of ammonia and titratable acid in nephritis. *J. Clin. Invest.* 2:255, 1926.

Widmer, B., Gerhardt, R. E., Harrington, J. T., and Cohen, J. J. Serum electrolyte and acid base composition: The influence of graded degrees of chronic renal failure. *Arch. Int. Med.* 139:1099, 1979.

Wrong, O., and Davies, H. E. F. The excretion of acid in renal disease. *Quart. J. Med.* 28:259, 1959.

Diabetic Ketoacidosis   Adrogué, H. J., Eknoyan, G., and Suki, W. N. Diabetic ketoacidosis: Role of the kidney in the acid-base homeostasis re-evaluated. *Kidney Int.* 25:591, 1984.

Adrogué, H. J., Wilson, H., Boyd, A. E., Suki, W. N., and Eknoyan, G. Plasma acid-base patterns in diabetic ketoacidosis. *N. Engl. J. Med.* 307:1603, 1982.

Atchley, D. W., Loeb, R. F., Richards, D. W., Benedict, E. M., and Driscoll, M. E. On diabetic acidosis: A detailed study of electrolyte balances following the withdrawal and reestablishment of insulin therapy. *J. Clin. Invest.* 12:297, 1933.

Cahill, G. F., Jr. Ketosis. *Kidney Int.* 20:416, 1981.

King, A. J., Cooke, N. J., McCuish, A., Clarke, B. F., and Kirby, B. J. Acid-base changes during treatment of diabetic ketoacidosis. *Lancet* 1:478, 1974.

Kreisberg, R. A. Diabetic ketoacidosis: New concepts and trends in pathogenesis and treatment. *Ann. Intern. Med.* 88:681, 1978.

Marliss, E. B., Ohman, J. L., Jr., Aoki, T. T., and Kozak, G. P. Altered redox state obscuring ketoacidosis in diabetic patients with lactic acidosis. *N. Engl. J. Med.* 283:978, 1970.

Witters, L. A. Endocrine and metabolic emergencies. In H. L. May (ed.), *Emergency Medicine.* New York: Wiley, 1984. Chap. 27.

Lactic Acidosis   Arieff, A. I., Leach, W., Park, R., and Lazarowitz, V. C. Systemic effects of NaHCO₃ in experimental lactic acidosis in dogs. *Am. J. Physiol.* 242 (Renal Fluid Electrolyte Physiol. 11):F586, 1982.

Arieff, A. I., Park, R., Leach, W. J., and Lazarowitz, V. C. Pathophysiology of experimental lactic acidosis in dogs. *Am. J. Physiol.* 239 (Renal Fluid Electrolyte Physiol. 8):F135, 1980.

Graf, H., Leach, W., and Arieff, A. I. Metabolic effects of sodium bicarbonate in hypoxic lactic acidosis in dogs. *Am. J. Physiol.* 249 (Renal Fluid Electrolyte Physiol. 18):F630, 1985.

Kreisberg, R. A. Pathogenesis and management of lactic acidosis. *Ann. Rev. Med.* 35:181, 1984.

Madias, N. E. Lactic acidosis. *Kidney Int.* 29:752, 1986.

Oliva, P. B. Lactic acidosis. *Am. J. Med.* 48:209, 1970.

Park, R., and Arieff, A. I. Treatment of lactic acidosis with dichloroacetate in dogs. *J. Clin. Invest.* 70:853, 1982.

Waters, W. C., Hall, J. D., and Schwartz, W. B. Spontaneous lactic acidosis. The nature of the acid-base disturbance and considerations in diagnosis and management. *Am. J. Med.* 35:781, 1963.

Ethylene Glycol

Gennari, F. J. Serum osmolality. Uses and limitations. *N. Engl. J. Med.* 301:102, 1984.

Levinsky, N. G. (Discussant). Case records of the Massachusetts General Hospital. *N. Engl. J. Med.* 301:650, 1979.

Peterson, C. D., Collins, A. J., Himes, J. M., Bullock, M. L., and Keane, W. F. Ethylene glycol poisoning. Pharmacokinetics during therapy with ethanol and hemodialysis. *N. Engl. J. Med.* 304:21, 1981.

Diarrhea

Kildeberg, P. Metabolic acidosis in infantile gastroenteritis. *Acta Paediatr. Scand.* 54:155, 1965.

Watten, R. H., Morgan, F. M., Songkhla, Y. N., Vanikiati, B., and Phillips, R. A. Water and electrolyte studies in cholera. *J. Clin. Invest.* 38:1879, 1959.

Renal Tubular Acidosis

Arruda, J. A. L., and Kurtzman, N. A. Mechanisms and classification of deranged distal urinary acidification. *Am. J. Physiol.* 239 (Renal Fluid Electrolyte Physiol. 8):F515, 1980.

Brenes, L. G., Brenes, J. N., and Hernandez, M. M. Familial proximal renal tubular acidosis. A distinct clinical entity. *Am. J. Med.* 63: 244, 1977.

Buckalew, V. M., Jr., Purvis, M. L., Shulman, M. G., Herndon, C. N., and Rudman, D. Hereditary renal tubular acidosis. *Medicine* 53:229, 1974.

DuBose, T. D., Jr., and Caflisch, C. R. Validation of the difference in urine and blood carbon dioxide tension during bicarbonate loading as an index of distal nephron acidification in experimental models of distal renal tubular acidosis. *J. Clin. Invest.* 75:1116, 1985.

Gennari, F. J., and Cohen, J. J. Renal tubular acidosis. *Ann. Rev. Med.* 29:521, 1978.

Goodman, A. D., Lemann, J., Jr., Lennon, E. J., and Relman, A. S. Production, excretion and net balance of fixed acid in patients with renal acidosis. *J. Clin. Invest.* 44:495, 1965.

Halperin, M. J., Goldstein, M. B., Haig, A., Johnson, M. D., and Steinbaugh, B. J. Studies on the pathogenesis of Type I (distal) renal tubular acidosis as revealed by the urinary P_{CO_2} tensions. *J. Clin. Invest.* 53:669, 1974.

Kurtzman, N. A. Acquired distal renal tubular acidosis. *Kidney Int.* 24:807, 1983.

Lightwood, R., Payne, W. W., and Black, J. A. Infantile renal acidosis. *Pediatrics* 12:628, 1953.

McSherry, E. Renal tubular acidosis in childhood. *Kidney Int.* 20:799, 1981.

Morris, R. C., Jr., and Sebastian, A. Renal Tubular Acidosis and the Fanconi Syndrome. In J. B. Stanbury, J. B. Wyngaarden, D. S. Frederickson, J. L. Goldstein, M. S. Brown (eds.). *The Metabolic Basis of Inherited Disease* (5th ed.). New York: McGraw-Hill, 1983.

Reynolds, T. B. Observations on the pathogenesis of renal tubular acidosis. *Am. J. Med.* 25:503, 1958.

Sebastian, A., McSherry, E., and Morris, R. C., Jr. Renal potassium wasting in renal tubular acidosis (RTA). Its occurrence in types 1 and 2 RTA despite sustained correction of systemic acidosis. *J. Clin. Invest.* 50:667, 1971.

Other Causes Cohen, J. J. Methylmalonic acidemia. *Kidney Int.* 15:311, 1978.

Gennari, F. J. Acid-base balance in dialysis patients. *Kidney Int.* 28:678, 1985.

Gonda, A., Gault, H., Churchill, D., and Hollomby, D. Hemodialysis for methanol intoxication. *Am. J. Med.* 64:749, 1978.

Levy, L. J., Duga, J., Girgis, M., and Gordon, E. E. Ketoacidosis associated with alcoholism in nondiabetic subjects. *Ann. Intern. Med.* 78:213, 1973.

McCoy, H. G., Cipolle, R. J., Ehlers, S. M., Sawchuk, R. J., and Zaske, D. E. Severe methanol poisoning. Application of a pharmacokinetic model for ethanol therapy and hemodialysis. *Am. J. Med.* 67:804, 1978.

Sebastian, A., Schambelan, M., Lindenfeld, S., and Morris, R. C., Jr. Amelioration of metabolic acidosis with fludrocortisone therapy in hyporeninemic hypoaldosteronism. *N. Engl. J. Med.* 297:576, 1977.

5 : Metabolic Alkalosis: Clinical Examples

Causes

The causes of metabolic alkalosis are listed in Table 5-1. Note that the conditions are divided into two major categories, based on whether or not the alkalosis can be corrected by giving Cl^-.

Table 5-1 : <u>Causes of metabolic alkalosis</u>
In order to emphasize the causative role of deficits in Cl^- in some of the disturbances, the list is divided into those disorders that respond to the administration of Cl^- and those that do not.*

<u>Responsive to Chloride:</u>
  Vomiting or gastric drainage
  Diuretic therapy with agents that inhibit NaCl reabsorption
    Furosemide
    Ethacrynic acid
    Metolazone
    Bumetanide
    Thiazides
  Sudden relief from chronic hypercapnia (posthypercapnic alkalosis)
  Certain types of diarrhea that lead to predominant loss of Cl^- (congenital chloridorrhea)
  Administration of alkali, especially in presence of ECF volume contraction or renal insufficiency
    HCO_3^-
    Salts of organic acids oxidized to yield HCO_3^- (e.g., lactate, citrate, acetate)
    Milk-alkali syndrome
<u>Resistant to Chloride:</u>
  Excess mineralocorticoids
    Cushing's syndrome
    Hyperaldosteronism
    ACTH-secreting tumors
    Licorice ingestion (mimics hyperaldosteronism)
  Bartter's syndrome
  Severe K^+ depletion (K^+ <2.0 meq/L)

*Administration of alkali is included under the category that is responsive to Cl^- because concomitant Cl^- intake will reduce serum $[HCO_3^-]$.

Prolonged Vomiting

A 44-year-old man had had a duodenal ulcer for at least 8 years. Once previously he had bled from the ulcer, but he had refused an operation. For the last 2 weeks he had been vomiting intermittently, and during the past several days he had kept down very little food or liquid. He had lost approximately 3 kg in weight, and on admission he was vomiting and quite weak.

On physical examination, his mucous membranes were dry, and he had poor skin turgor. Blood pressure was 110/70 mm Hg, pulse 92 per minute and regular, and respiratory rate 15 per minute. He had no peripheral edema. His abdomen was distended and tympanitic, and there was tenderness in the epigastrium. He had hypoactive deep tendon reflexes. Selected laboratory values, on admission and 48 hours later, are shown in Table 5-2.

Comment

The arterial blood gas analysis on admission indicated the presence of a metabolic alkalosis. The pH was in the alkalemic range (>7.45) and [HCO_3^-] was increased. In order to evaluate whether the ventilatory response (i.e., the rise in PCO_2; Table 3-2) is consistent with that seen in uncomplicated metabolic alkalosis, one can use the rule of thumb (Table 3-3) or the confidence bands (Fig. 3-2). Applying the former technique: [HCO_3^-] has increased from a normal value of 24 mmoles per liter to 45, i.e., by 21 mmoles per liter; since PCO_2 should increase, on average, by 0.7 mm Hg for each mmole per liter rise in [HCO_3^-], the PCO_2 should have risen by approximately 21 $\times$ 0.7, or 15 mm Hg. The new PCO_2 would therefore be 40 + 15, or 55 mm Hg, and the observed response of 56 mm Hg is thus consistent with an uncomplicated metabolic alkalosis.

The venous values on admission demonstrated hypokalemia, a [Cl^-] that was lowered proportionally much more than the [Na^+], and a very high total CO_2 content (as expected, some 2 mmoles per liter higher than the arterial [HCO_3^-]; see Laboratory Tests, Chap. 3). The anion gap was increased very slightly. This increase, which is consistent with moderately severe un-

Table 5-2 : Results of laboratory tests in a 44-year-old man before and after treatment for metabolic alkalosis due to prolonged vomiting
The numbers in parentheses indicate estimated values.

Test	On admission	48 Hours after admission
Arterial blood:		
pH	7.53	7.45
PCO_2 (mm Hg)	56	45
[H^+] (nmoles/L)	30 (30)	35 (35)
[HCO_3^-] (mmoles/L)	45 (45)	30 (31)
Venous serum:		
[Na^+] (mmoles/L)	135	141
[K^+] (mmoles/L)	2.7	3.4
[Cl^-] (mmoles/L)	70	97
Total CO_2 (mmoles/L)	47	32
Anion gap (meq/L)	18	12

complicated metabolic alkalosis, is due partly to a pH-dependent rise in the number of negative charges on plasma proteins (Figs. 1-5 and 3-3), and in part to the elevation of plasma protein concentration that occurs with contraction of the extracellular fluid volume (ECF).

Pathogenesis

The pathogenesis of metabolic alkalosis — not only that resulting from vomiting or nasogastric drainage, but also metabolic alkaloses arising from other causes (see below) — is best understood by dividing the events into an initiating phase, during which the alkalosis is generated, and a maintenance phase, during which it is perpetuated.

GENERATION. The initiating event, with vomiting and nasogastric drainage, is the loss of HCl from the body (Figs. 5-1A and B). As Figure 5-1A shows, secretion of H^+ into the stomach results in addition of HCO_3^- to the body fluids, just as it does in the renal tubules. Normally, the HCl that is secreted into the stomach traverses into the duodenum, where it stimulates the secretion of an equivalent amount of $NaHCO_3$. When the gastric HCl is eliminated from the body by vomiting or nasogastric drainage, however, $NaHCO_3$ is not secreted into the duodenum but is retained in the plasma (Fig. 5-1A). Thus, loss of acidic gastric contents produces a change in plasma composition, which is characterized by a decrease in $[H^+]$ (i.e., alkalosis), an increase in $[HCO_3^-]$, and a decrease in $[Cl^-]$ (Fig. 5-1B). By contrast, losses of sodium and potassium are small; when gastric fluid is maximally acid (pH = 1.0), each liter contains approximately 100 mmoles of HCl, 15 mmoles of NaCl, and 10 mmoles of KCl (Table 1-3).

MAINTENANCE. Once the alkalosis has been established by gastrointestinal losses of HCl, it is maintained by an increase in the renal reabsorption of HCO_3^-, which in turn is induced by depletion of chloride and contraction of the extracellular fluid volume (ECF). It was pointed out in Chapter 1 (under Utilization of the Various Buffers. Addition of Strong Alkali) that normally the kidneys excrete excess $NaHCO_3$ very rapidly. If, however, losses of chloride induced by vomiting or nasogastric drainage are not replaced, renal handling of alkali is altered by the following sequence of events (Fig. 5-1B).

The changes in the composition of plasma that occur in the initiating phase are necessarily paralleled by a modification of the glomerular filtrate; i.e., the amount of filtered HCO_3^- is increased and that of filtered Cl^- is decreased (Fig. 5-1B). At first, these changes lead to an increased urinary excretion of HCO_3^-, which is accompanied by Na^+ (interrupted arrow in Fig. 5-1B).

118

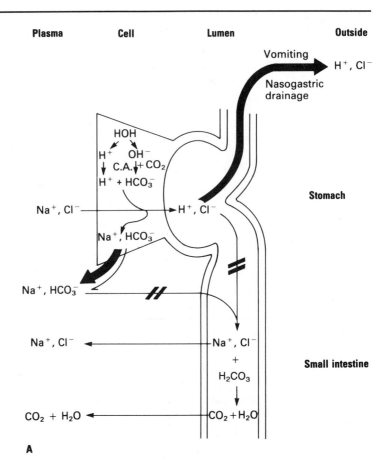

A

Fig. 5-1 :
A. Mechanism by which vomiting or nasogastric drainage leads to net addition
of NaHCO₃ to the plasma. The thin arrows denote normal pathways for the gastric secretion of HCl; the heavy lines, the altered routes that occur during
vomiting or nasogastric drainage. C.A. = Carbonic anhydrase.

Generation Phase

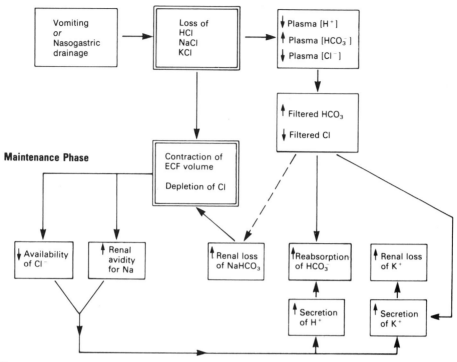

Maintenance Phase

B

B. Pathogenesis of chloride-depletion metabolic alkalosis caused by vomiting
or nasogastric drainage. The sequence of events is described in the text. The
two large blocks, framed in double lines, denote the major factors that initiate
and then maintain the alkalosis. The interrupted arrow indicates that renal loss
of NaHCO$_3$ occurs initially but does not continue once contraction of the extracellular fluid (ECF) volume and depletion of Cl$^-$ are established. Not shown
in the figure is a decrease in GFR, which, when present, contributes to the
maintenance of the alkalosis by reducing renal loss of bicarbonate.

But then the urinary loss of Na^+, coupled with the loss of NaCl and water from the stomach, eventuates in contraction of the extracellular fluid volume and consequent avid reabsorption of Na^+. All reabsorbed Na^+ must be either accompanied by an anion (mostly Cl^-) or balanced electrically by secreted H^+ or K^+. Inasmuch as delivery of Cl^- to the tubular system (the filtered load) is greatly curtailed, the first mechanism is reduced while the second is increased, so that the secretion of H^+ and K^+ is enhanced; and since the kidneys either reabsorb or generate a HCO_3^- for every H^+ that the tubules secrete (Figs. 2-1, 2-3, and 2-6), the high plasma $[HCO_3^-]$ is sustained. Thus, selective depletion of Cl^- and contraction of the extracellular fluid are the key factors that maintain the alkalosis: Increased renal avidity for Na^+ in the face of a low filtered load of Cl^- results in increased secretion of H^+ and consequent increased reabsorption of HCO_3^-. This change persists until Cl^- is replaced, so that this ion can again accompany the reabsorbed Na^+.

The above-described events represent the current explanation for the generation and maintenance of metabolic alkalosis due to loss of gastric juice. Neither volume contraction nor moderate K^+ depletion alone causes metabolic alkalosis, at least in humans. Similarly, although secretion of aldosterone is stimulated by volume contraction, there is no evidence that abnormally high plasma levels of aldosterone are required to maintain this type of alkalosis. Rather, depletion of Cl^- coupled with even only a slight contraction of the ECF appear to be the only requirements to produce and sustain this form of metabolic alkalosis (see below, Administration of a Diuretic).

Even though K^+ depletion appears not to be a causative factor, it nevertheless is almost invariably present in this type of metabolic alkalosis. The reason is not only the loss of KCl in the vomitus but, more important, the increased secretion of K^+ into the renal tubules, and hence its increased urinary excretion.

Treatment

The cause of gastric fluid loss obviously should be corrected, if at all possible. But often in these situations the acid-base disorder must be treated first, in order to get the patient into better shape for an operation. In this patient the vomiting was relieved through insertion of a nasogastric tube, and secretion of gastric acid was blocked with cimetidine (a histamine receptor antagonist). Simultaneously, parenteral therapy was begun, as outlined below.

Appropriate treatment follows logically from an understanding of pathogenesis. The key is to replenish Cl^- stores, so that renal secretion of hydrogen ion can be reduced. When sufficient Cl^-

has been given so that its plasma concentration and hence its filtered load (GFR $\times$ P_{Cl^-}) are increased, Cl^- can again accompany the reabsorbed Na^+. At that point, secretion of H^+ will decrease toward the normal level, excessive reabsorption of HCO_3^- will therefore stop, and the extra HCO_3^- aboard will be excreted. Consequently, the alkalosis will be corrected. Note that this therapeutic regimen will also minimize further K^+ losses, since the secretion not only of H^+ but also of K^+ will lessen.

In practice, one gives the required Cl^- as the potassium and sodium salts, in order to replenish K^+ stores and to restore ECF volume, respectively. Correction of ECF volume contraction will eliminate renal Na^+ avidity and speed recovery. The present patient received 5 liters of isotonic saline (Fig. 1-8A), with 40 mmoles of KCl added to each liter (Fig. 1-8C), over 48 hours. By the end of that period, the laboratory values had returned to near-normal levels (Table 5-2), and within 2 to 3 more days they were completely normal. The patient then underwent surgery for correction of an obstructing duodenal ulcer.

Administration of a Diuretic

A 36-year-old man with the nephrotic syndrome was treated for extensive edema with dietary salt restriction and with diuretics. Initial therapy with hydrochlorothiazide did not clear the edema, and he was therefore given a more efficacious diuretic, furosemide. The dosage of this drug was increased to the very high level of 320 mg per day, and after one month on daily furosemide the patient was free of edema. However, by that time the patient felt weak and "washed out," both figuratively and literally. He had not experienced nausea, vomiting, or diarrhea. He had adhered strictly to the low salt diet, and he had not used a salt substitute. Physical examination was unremarkable; there was no edema, and the blood pressure was 110/70 mm Hg and did not change with posture.

Comment

The history — with continuous administration of a very potent diuretic given in high dosage — points to diuretic-induced metabolic alkalosis and K^+ deficiency. The initial laboratory tests on venous serum (Table 5-3) supported these impressions. Not only was the total CO_2 elevated, but the [Cl^-] was disproportionately low in relation to [Na^+], which follows from the pathogenesis for this type of metabolic alkalosis (discussed below). These findings, coupled with the low [K^+], are so consistent with the diagnosis that an arterial blood gas analysis was probably unnecessary in this patient. Nevertheless, that procedure was carried out, and the pH of 7.48 clearly ruled out the possibility that the high total CO_2 in the venous sample reflected a respiratory acidosis (Table 3-2). Given the rise in arterial [HCO_3^-] of 10

Table 5-3 : **Results of laboratory tests in a 36-year-old man with nephrotic syndrome treated with furosemide**
The numbers in parentheses indicate estimated values.

Test	Initial evaluation	After treatment
Arterial blood:		
pH	7.48	
P_{CO_2} (mm Hg)	47	
$[H^+]$ (nmoles/L)	33 (33)	
$[HCO_3^-]$ (mmoles/L)	34 (34)	
Venous serum:		
$[Na^+]$ (mmoles/L)	135	136
$[K^+]$ (mmoles/L)	2.9	3.3
$[Cl^-]$ (mmoles/L)	86	95
Total CO_2 (mmoles/L)	36	29
Anion gap (meq/L)	13	12

mmoles per liter (from 24 to 34), one would anticipate an increase in P_{CO_2} of 10 × 0.7, or 7 mm Hg, using the rule of thumb given in Table 3-3. The identity of predicted with measured P_{CO_2} of 47 mm Hg thus indicates an uncomplicated metabolic alkalosis.

Pathogenesis

GENERATION. The initiating phase of this type of metabolic alkalosis probably has at least three components: (1) loss of chloride, (2) stimulation of Na^+/H^+ exchange (and hence of HCO_3^- reabsorption; Fig. 2-1) in the loops of Henle, and (3) stimulation of H^+ secretion (and hence of HCO_3^- reabsorption and acid excretion) in nephron segments beyond the loops.

Furosemide is one of several potent diuretic agents (Table 5-1) that inhibit the reabsorption of NaCl from the thick ascending limbs of Henle; it does so by inhibiting a so-called co-transporter in the luminal (or apical) membrane (for orientation, see Fig. 2-1), which carries Na^+ and Cl^- into the cell. Inhibition of this co-transporter thus lowers the intracellular $[Na^+]$, a process that is abetted because Na^+ continues to be transported out of the cell at the opposite, basolateral membrane via an active Na^+/K^+-ATPase pump (Fig. 2-1). The decrease in intracellular $[Na^+]$ promotes the passive entry of some Na^+ into the cell without Cl^- and thereby stimulates the mechanism that exchanges H^+ for Na^+ (Fig. 2-1). The net effect of furosemide on the loops of Henle is thus decreased reabsorption of Na^+ and Cl^-, and increased secretion of H^+ and reabsorption of HCO_3^-.

Although most of the NaCl that escapes reabsorption in the loops of Henle because of furosemide is excreted in the urine (thereby causing the diuresis), some is reabsorbed in nephron segments beyond the loops. This distal reabsorption is enhanced in edematous states (such as the nephrotic syndrome or congestive heart failure), where the generalized edema results from an inability of the kidneys to excrete Na^+ appropriately. Consequently, in a patient such as the present one, who exhibits the nephrotic syndrome and is being treated with furosemide, we have the combination of increased delivery of Na^+ and Cl^- (and ultimately of water) to the distal tubules and collecting ducts plus a stimulus for avid Na^+ reabsorption. This combination is known to result in increased secretion not only of K^+ but also of H^+, and the latter is accompanied, mole per mole, by either reabsorption of filtered HCO_3^- or generation of HCO_3^- through acid excretion (Fig. 2-8). Thus, events in the more distal portions of the nephron, which are secondary to the influence of furosemide on the loops of Henle, result in a preferential loss of H^+ and Cl^- from the body. In this manner the pathogenesis of diuretic-induced metabolic alkalosis is similar to that resulting from loss of gastric contents.

MAINTENANCE. As in the case of gastric alkalosis, selective loss of H^+ and Cl^- leads to depletion of Cl^- and a disproportionate reduction in plasma $[Cl^-]$ (Table 5-3). The resultant reduction in the filtered load of Cl^- (GFR $\times$ P_{Cl^-}) plus renal avidity for Na^+ result in a sustained increase of H^+ secretion and HCO_3^- reabsorption, so that the alkalosis is perpetuated (Fig. 5-1B); only here the increased avidity for Na^+ is an effect of the nephrotic syndrome on the kidneys, and is not secondary to contraction of the extracellular fluid volume. That volume is in fact expanded when there is generalized edema. Potassium depletion and hypokalemia are characteristic of diuretic-induced metabolic alkalosis (Table 5-3), partly because K^+ reabsorption in the loops of Henle is inhibited by furosemide, and partly because K^+ secretion in the distal and collecting tubules is stimulated by the increase in the delivery of Na^+ and water.

The diuretic agents that can induce metabolic alkalosis are listed in Table 5-1. All inhibit the reabsorption of NaCl from the thick ascending limb of Henle's loop (or from the early distal tubule, in the case of thiazides), and cause a disproportionate loss of Cl^- in the urine, as described above. The metabolic alkalosis induced by these diuretic agents is usually milder than that seen with losses from the upper gastrointestinal tract. The most likely explanation for this difference is that the losses of Cl^- induced by diuretics are limited by countervailing forces within the kidneys (e.g., a fall in glomerular filtration rate due to eventual volume contraction and tubuloglomerular feedback, and a consequent increase in proximal Cl^- reabsorption driven by enhanced proximal Na^+ reabsorption), whereas the losses of Cl^- in gastric fluid are unopposed.

Treatment

Treatment for this type of patient poses a dilemma: Because the alkalosis responds to Cl^- (Table 5-1), administration of Cl^- will correct the acid-base abnormality; if, however, the Cl^- is given as the sodium salt, the avidity of the kidneys for Na^+ in edematous states will lead to reappearance of edema. Therefore, the Cl^- should be administered as the potassium salt. Although this measure will lessen the alkalosis and the K^+ deficiency, a return to normal plasma $[HCO_3^-]$ and $[K^+]$ is unlikely to occur as long as diuretic therapy is continued. In the present patient the nephrotic syndrome was unresponsive to other therapy, and the diuretic therefore had to be continued in order to prevent recurrence of edema.

Correction of mild metabolic alkalosis is not always necessary. A more important therapeutic goal is to increase plasma $[K^+]$ to above 3.0 mmoles per liter, a level at which fewer symptoms (such as weakness) are present. Therefore, the patient was given KCl, 40 mmoles per day, and he was encouraged to eat foods rich in K^+. At the same time the dosage of furosemide was gradually reduced, while his body weight was monitored closely. Several weeks later the serum electrolytes had improved considerably, albeit only partially (Table 5-3), and the patient was feeling better.

**Hyperaldo-
steronism**

The patient is a 45-year-old man who, on routine physical examination, was found to have a high blood pressure of 170/100 mm Hg as the only abnormal finding. He had no symptoms whatsoever. Five years earlier, at the time of a similar examination, the patient's blood pressure had been normal. He was taking no medications, was following a normal diet, and had neither lost nor gained weight over the previous year. As part of an evaluation for hypertension, the routine electrolytes listed in Table 5-4 were obtained.

**Table 5-4 : Results of laboratory tests
in a 45-year-old man with hypertension**
The numbers in parentheses indicate estimated values.

Test	Initial evaluation	After surgery
Arterial blood:		
pH	7.50	
P_{CO_2} (mm Hg)	47	
$[H^+]$ (nmoles/L)	32 (32)	
$[HCO_3^-]$ (mmoles/L)	35 (35)	
Venous serum:		
(Na^+) (mmoles/L)	148	142
$[K^+]$ (mmoles/L)	2.8	4.0
$[Cl^-]$ (mmoles/L)	100	103
Total CO_2 (mmoles/L)	37	26
Anion gap (meq/L)	11	13

Comment

Here is one instance where routine laboratory tests — i.e., tests ordered without a specific prior suspicion of the diagnosis — provided the key clue. As in the two patients already discussed in this chapter, the combination of an elevated total CO_2 content and a marked reduction in plasma $[K^+]$ strongly suggested the diagnosis of metabolic alkalosis. The arterial measurements confirmed this impression: The pH was in the alkalemic range and the elevation of PCO_2 was consistent with the response seen in uncomplicated metabolic alkalosis (Table 3-3).

In contrast to the previous patients, however, this patient showed no evidence of extracellular fluid volume (ECF) contraction or edema. His body weight had been stable and he was hypertensive. Nor was there any history suggesting abnormal losses of Cl^-, and the plasma $[Cl^-]$ was in the normal range. Depletion of Cl^- as a cause of the alkalosis was further ruled out by measuring the urinary concentration of Cl^-, which was 63 mmoles per liter. This result showed that delivery of Cl^- to the distal tubules and collecting ducts must have been sufficient to correct any chloride-responsive alkalosis — yet, a high arterial pH and $[HCO_3^-]$ persisted.

Thus, the differential diagnosis for the cause of the alkalosis lay in the conditions listed in Table 5-1 as being resistant to Cl^-. With hypertension as the predominant sign, primary hyperaldosteronism emerged as the most likely possibility. In this condition, aldosterone is secreted continuously and autonomously by abnormal adrenal tissue, as in hyperplasia or an adenoma of the adrenal glands. The diagnosis was confirmed by measuring a high aldosterone and a low renin level in plasma, as well as by locating an adrenal adenoma on CT (computerized tomography) scan.

Pathogenesis

In contrast to the Cl^--dependent (or Cl^--responsive) forms of metabolic alkalosis, in this type (Cl^--resistant) the high rates of renal HCO_3^- reabsorption (occurring pari passu with H^+ secretion — Figs. 2-1, 2-3, and 2-6) and K^+ secretion occur independently of the renal handling of Cl^-. Rather, they result directly or indirectly, from the action(s) of adrenal mineralocorticoids. As will be discussed below, the initiating event for this disorder is a sustained increase in aldosterone secretion coupled with sufficient sodium intake to insure adequate sodium delivery to the distal tubules and collecting ducts.

Although many questions remain concerning the way in which mineralocorticoids stimulate the secretion of H^+ and K^+, the following represents our current understanding of the mechanisms involved. Basic to starting the chain of events is the action of aldosterone to stimulate Na^+/K^+-ATPase in the peritubular (basolateral; see Fig. 2-1) membrane of cells in the distal tubules and collecting ducts. This effect stimulates the transport of Na$^+$ out of cells and K$^+$ into them. As a result, the intracellular concentration of K$^+$ increases, facilitating passive K$^+$ secretion into the tubular lumen. In addition, there is evidence that at least two other effects of aldosterone — namely, increased luminal membrane permeability for K$^+$ and a decrease in the electrical potential difference across the luminal membrane — also aid in the secretion of K$^+$ and hence its urinary excretion.

The other consequence of the effect of aldosterone on the basolateral pump — namely, that of transporting Na$^+$ out of cells — is to lower the intracellular concentration of Na$^+$. This change provides the "sink," so to speak, for increased passive entry of Na$^+$ from tubular lumen into the cell, a process that is abetted by a further effect of aldosterone that increases the permeability of the luminal membrane for Na$^+$. As a result of Na$^+$ entry into the cell from the tubular lumen, the electrical potential difference across the luminal membrane decreases, and this change, in turn, stimulates secretion of H$^+$ into the lumen. In addition, aldosterone can enhance H$^+$ secretion independently of Na$^+$ movement, possibly by directly stimulating the energy-consuming H$^+$ pumps in the luminal membrane.

Consistent with a Na$^+$-dependent component for K$^+$ and H$^+$ secretion is the observation that excess mineralocorticoids manifest their effect on plasma [K$^+$] and [HCO$_3^-$] only when the dietary intake of Na$^+$ is sufficient to insure adequate delivery of Na$^+$ to more distal parts of the nephron (i.e., only when moderate amounts of Na$^+$ are excreted in the urine). Initially, aldosterone causes a brief period of Na$^+$ retention (3 to 4 days), during which there is a slight gain in body weight, signalling an expansion of ECF volume. Then the so-called escape from the Na$^+$-retaining effects of aldosterone occurs, so that a new steady state is attained, in which Na$^+$ is again excreted appropriately in the urine despite continued high plasma levels of aldosterone. The escape is thought to result from inhibition of Na$^+$ reabsorption in proximal tubules and perhaps in the loops of Henle, and therefore from delivery of more Na$^+$ to later portions of the nephron and into the urine. The transfer of a greater Na$^+$ load to the distal nephron leads to increased secretion not only of K$^+$ but also of H$^+$ — albeit not by a direct exchange mechanism.

An additional difference between Cl$^-$-depletion alkalosis and the mineralocorticoid-induced variety is that in the latter the severity of the alkalosis appears to correlate with the degree of K$^+$ depletion. Although the reasons for this phenomenon are not fully understood, it is a common clinical finding, illustrated in the present patient, that the alkalosis is much more severe in the presence of hypokalemia (Table 5-4) than can be induced in normal subjects with aldosterone only.

In summary, then, increased H^+ secretion (and hence HCO_3^- reabsorption) in the presence of excess mineralocorticoids results in part from increased delivery of Na^+ to the distal nephron, and in part from processes that are independent of Na^+. Simultaneously, secretion of K^+ is stimulated, and when this effect has led to K^+ depletion, the metabolic alkalosis is aggravated.

Treatment

The most rational approach to therapy is to correct the primary abnormality — in this instance, to remove the abnormal adrenal tissue. It is wise, however, first to correct the acid-base abnormality so the patient will be in a better condition for surgery. This goal was accomplished for this patient simply by administration of K^+, which (in conformity with the importance of K^+ depletion in this type of alkalosis) usually suffices to ameliorate the acid-base picture; alternatively, the aldosterone antagonist spironolactone could have been given. Thereafter, an adenoma above the right kidney was removed, and with this measure all the abnormalities in the plasma were corrected (Table 5-4).

Bartter's Syndrome

A 28-year-old woman sought medical attention because of mild weakness and fatigue. These symptoms had been present for several years, but they had become more troublesome during the recent months. She reported no other medical problems, except for mild depression and some concern about gaining weight. As a result, her dietary intake had been sparse and erratic. She denied taking laxatives, diuretics, or any other medications, save for an occasional aspirin, and she stated that she had not vomited or had diarrhea.

Physical examination was negative except for hypoactive deep tendon reflexes. Her weight was 48 kg, her height 156 cm, and her blood pressure 110/70 mm Hg while she was sitting and 96/60 when she stood up. Results of laboratory tests are shown in Table 5-5.

Comment

The laboratory values indicated an uncomplicated metabolic alkalosis with some selective lowering of the plasma $[Cl^-]$, and hypokalemia. In contrast to the last patient, this one was not hypertensive; rather, she had some postural hypotension, which suggested contraction of the extracellular fluid volume. Thus, this patient posed the question of whether a Cl^--sensitive or a Cl^--resistant metabolic alkalosis was present (Table 5-1). Although her history would seem to have excluded the Cl^--sensitive causes, the possibility had to be considered that the

Table 5-5 : Results of laboratory tests
in a 28-year-old woman with Bartter's syndrome
The numbers in parentheses are estimated values.

Test	Initial evaluation	After treatment
Arterial blood:		
pH	7.52	
P_{CO_2} (mm Hg)	48	
[H^+] (nmoles/L)	30 (31)	
[HCO_3^-] (mmoles/L)	38 (37)	
Venous serum:		
[Na^+] (mmoles/L)	140	142
[K^+] (mmoles/L)	2.5	3.4
[Cl^-] (mmoles/L)	85	96
Total CO_2 (mmoles/L)	40	32
Anion gap (meq/L)	15	14

patient was a surreptitious vomiter or an abuser of diuretics — practices that are sometimes resorted to by patients with personality disorders, especially if they are concerned about being overweight. Such patients typically deny vomiting or taking diuretics, and uncovering these facts, when they are present, requires close observation and often the help of the patient's family or friends.

Measurement of the urinary Cl^- concentration is a simple test, which can help to rule out surreptitious vomiting (although not abuse of diuretics, as explained below). If the urinary [Cl^-] is relatively high (>30 mmoles per liter), then vomiting is excluded as a cause of the alkalosis, since (as explained earlier in this chapter) the selective loss of Cl^- during vomiting leads to avid renal conservation of Cl^-. This test, however, cannot be applied to exclude diuretic abuse, since the diuretics increase urinary excretion of Cl^-. In either case — surreptitious vomiting or surreptitious use of diuretics — one can apply a therapeutic test of administering NaCl and KCl (see above under Prolonged Vomiting, Treatment). If the alkalosis is corrected, then it was Cl^--sensitive; if it persists, then one of the Cl^--resistant forms of metabolic alkalosis (Table 5-1) is likely to be involved.

In the present patient, the urinary concentration of Cl^- was 53 mmoles per liter despite the persistence of alkalosis. Careful probing of the history failed to reveal evidence of vomiting or of diuretic abuse; therefore, a therapeutic test with Cl^- salts was deemed to be unnecessary. Instead, one of the causes for Cl^--resistant metabolic alkalosis (Table 5-1) was sought. The patient showed no evidence of excess mineralocorticoid secretion. She had not ingested licorice in large amounts, and she was not sufficiently depleted of K^+ to have developed the alkalosis on that basis. Thus, Bartter's syndrome became the main possibility, and this suspicion was supported by the finding of an elevated plasma renin activity and aldosterone concentration, as well as by a renal biopsy, which showed hypertrophy and hyperplasia of the juxtaglomerular cells. Continued loss of Cl^- in the urine, in the face of evidence for a normal hormonal response to volume contraction, is characteristic of Bartter's syndrome.

Pathogenesis

Bartter's syndrome is a rare disorder, which is seen mainly in children and young adults. As was true in the present instance, the patients usually seek medical attention because of somewhat vague symptoms related to hypokalemia — i.e., weakness and fatigue. They are normotensive or may even have a slightly low blood pressure because of mild volume depletion.

Although the pathogenesis of the syndrome is not yet clearly defined, current evidence suggests that faulty reabsorption of NaCl from the thick ascending limbs of Henle may be the primary defect. If that hypothesis is correct, then the metabolic alkalosis and hypokalemia in Bartter's syndrome result from the same chain of events as described above for administration of diuretics that act on the loop of Henle — except that in Bartter's syndrome the stimulus for avid Na^+ reabsorption in distal tubules and collecting ducts arises from volume contraction rather than from a renal disorder, as in the nephrotic syndrome. But even though the pathogenesis is analogous in the two conditions, the alkalosis of Bartter's syndrome cannot be corrected through administration of Cl^- salts, because any administered salts continue to be rapidly excreted in the urine. Because of the volume depletion, patients with Bartter's syndrome have elevated plasma levels of renin, aldosterone, and prostaglandins.

Treatment

It has been found empirically that inhibitors of prostaglandin synthesis, such as indomethacin, provide the most dramatic response. Therefore, they are now the mainstay of therapy. For reasons that are unclear, these agents correct the depletion of the extracellular fluid volume, and they lead to partial repair of the metabolic alkalosis and the K^+ depletion. Although some further correction might be achieved through a combination of KCl and drugs that inhibit renal secretion of K^+ (spironolactone, amiloride, triamterene), total correction of the alkalosis and hypokalemia is usually unnecessary. In most instances, the patient's symptoms can be eliminated with prostaglandin inhibitors alone. This was true of the present patient, whose laboratory values after treatment are shown in Table 5-5.

Milk-Alkali Syndrome

A clinical history of this disorder — another example of metabolic alkalosis — was presented as Problem 3-2 (p. 81) and is discussed in the answer to that problem at the end of the book (p. 183).

Selected References

General

Bone, J. M., Cowie, J., Lambie, A. T., and Robson, J. S. The relationship between arterial P_{CO_2} and hydrogen ion concentration in chronic metabolic acidosis and alkalosis. *Clin. Sci. Mol. Med.* 46:113, 1973.

Cohen, J. J., and Kassirer, J. P. *Acid-Base.* Boston: Little, Brown, 1982. Chap. 9.

Garella, S., Chang, B. S., and Kahn, S. I. Dilution acidosis and contraction alkalosis: Review of a concept. *Kidney Int.* 8:279, 1975.

Harrington, J. T. Metabolic alkalosis. *Kidney Int.* 26:88, 1984.

Maddox, D. A., and Gennari, F. J. Load-dependence of proximal tubular bicarbonate reabsorption in chronic metabolic alkalosis in the rat. *J. Clin. Invest.* 77:709, 1986.

Maddox, D. A., and Gennari, F. J. Proximal tubular bicarbonate reabsorption and P_{CO_2} in chronic metabolic alkalosis in the rat. *J. Clin. Invest.* 72:1385, 1983.

Madias, N. E., Adrogué, H. J., and Cohen, J. J. Maladaptive renal response to secondary hypercapnia in chronic metabolic alkalosis. *Am. J. Physiol.* 238 (Renal Fluid Electrolyte Physiol. 7):F283, 1980.

Madias, N. E., Ayus, J. C., and Adrogué, H. J. Increased anion gap in metabolic alkalosis. The role of plasma-protein equivalency. *N. Engl. J. Med.* 300:1421, 1979.

Madias, N. E., Bossert, W. H., and Adrogué, H. J. Ventilatory response to chronic metabolic acidosis and alkalosis in the dog. *J. Appl. Physiol.* 56:1640, 1984.

Schwartz, W. B., and Cohen, J. J. The nature of the renal response to chronic disorders of acid-base equilibrium. *Am. J. Med.* 64:417, 1978.

Chloride-Responsive Metabolic Alkalosis

Bleich, H. L., Tannen, R. L., and Schwartz, W. B. The induction of metabolic alkalosis by correction of potassium deficiency. *J. Clin. Invest.* 45:573, 1966.

Galla, J. H., Bonduris, D. N., Dumbauld, S. L., and Luke, R. G. Segmental chloride and fluid handling during correction of chloride-depletion alkalosis without volume expansion in the rat. *J. Clin. Invest.* 73:96, 1984.

Jacobson, H. R., and Seldin, D. W. On the generation, maintenance, and correction of metabolic alkalosis. *Am. J. Physiol.* 245 (Renal Fluid Electrolyte Physiol. 14):F425, 1983.
This article and the one by Luke and Galla below present two sides of a continuing controversy concerning the importance of extracellular fluid volume and potassium — as opposed to Cl^- — in the generation and repair of metabolic alkalosis.

Kassirer, J. P., Berkman, P. M., Lawrenz, D. R., and Schwartz, W. B. The critical role of chloride in the correction of hypokalemic alkalosis in man. *Am. J. Med.* 38:172, 1965.

Luke, R. G., and Galla, J. H. Chloride-depletion alkalosis with a normal extracellular fluid volume. *Am. J. Physiol.* 245 (Renal Fluid Electrolyte Physiol. 14):F419, 1983.

Schwartz, W. B., van Ypersele de Strihou, C., and Kassirer, J. P. Role of anions in metabolic alkalosis and potassium deficiency. *N. Engl. J. Med.* 279:630, 1968.

Gastrointestinal Loss of H+

Berger, B. E., Cogan, M. G., and Sebastian, A. Reduced glomerular filtration and enhanced bicarbonate reabsorption maintain metabolic alkalosis in humans. *Kidney Int.* 26:205, 1984.

Gamble, J. L., and Ross, S. G. The factors in the dehydration following pyloric obstruction. *J. Clin. Invest.* 1:403, 1925.

Kassirer, J. P., and Schwartz, W. B. Correction of metabolic alkalosis in man without repair of potassium deficiency. A re-evaluation of the role of potassium. *Am. J. Med.* 40:19, 1966.

Kassirer, J. P., and Schwartz, W. B. The response of normal man to selective depletion of hydrochloric acid. Factors in the genesis of persistent gastric alkalosis. *Am. J. Med.* 40:10, 1966.
These two important papers describe in detail the influence of gastric H+ loss on acid-base equilibrium under carefully controlled conditions, and they demonstrate the critical role of Cl− in the repair of this form of metabolic alkalosis.

McCallum, W. G., Lintz, J., Vermilye, H. N., Leggett, T. H., and Boas, E. The effect of pyloric obstruction in relation to gastric tetany. *Bull. Johns Hopkins Hosp.* 31:1, 1920.

Needle, M. A., Kaloyanides, G. J., and Schwartz, W. B. The effects of selective depletion of hydrochloric acid on acid-base and electrolyte equilibrium. *J. Clin. Invest.* 43:1836, 1964.

Wallace, M., Richards, P., Chesser, E., and Wrong, O. Persistent alkalosis and hypokalaemia caused by surreptitious vomiting. *Quart. J. Med.* 37:577, 1968.

Diuretics

Cannon, P. J., Heinemann, H. O., Albert, M. S., Laragh, J. H., and Winters, R. W. "Contraction" alkalosis after diuresis of edematous patients with ethacrynic acid. *Ann. Intern. Med.* 62:979, 1965.

Goldring, R. M., Cannon, P. J., Heinemann, H. O., and Fishman, A. P. Respiratory adjustment to chronic metabolic alkalosis in man. *J. Clin. Invest.* 47:188, 1968.

Hropot, M., Fowler, N., Karlmark, B., and Giebisch. G. Tubular action of diuretics: Distal effects on electrolyte transport and acidification. *Kidney Int.* 28:477, 1985.

Schwartz, W. B., and Wallace. W. M. Electrolyte equilibrium during mercurial diuresis. *J. Clin. Invest.* 30:1089, 1951.

Hyperaldosteronism

De Graeff, J., Struyvenberg, A., and Lameijer, L. D. F. The role of chloride in hypokalemic alkalosis. Balance studies in man. *Am. J. Med.* 37:778, 1964.

Kassirer, J. P., London, A. M., Goldman, D. M., and Schwartz, W. B. On the pathogenesis of metabolic alkalosis in hyperaldosteronism. *Am. J. Med.* 49:306, 1970.

Melby, J. C. Primary aldosteronism. *Kidney Int.* 26:769, 1984.

132

Bartter's Syndrome   Bartter, F. C., Pronove, P., Gill, J. R., Jr., and MacCardle, R. C. Hyperplasia of the juxtaglomerular complex with hyperaldosteronism and hypokalemic alkalosis. A new syndrome. *Am. J. Med.* 33:811, 1962.

Dunn, M. J. Prostaglandins and Bartter's syndrome. *Kidney Int.* 19:86, 1981.

Gill, J. R., Jr., and Bartter, F. C. Evidence for a prostaglandin-independent defect in chloride reabsorption in the loop of Henle as a proximal cause of Bartter's syndrome. *Am. J. Med.* 65:766, 1978.

Stein, J. H. The pathogenetic spectrum of Bartter's syndrome. *Kidney Int.* 28:85, 1985.

Verberckmoes, R., Van Damme, B., Clement, J., Amery, A., and Michielsen, P. Bartter's syndrome with hyperplasia of renomedullary cells: Successful treatment with indomethacin. *Kidney Int.* 9:302, 1976.

Other Causes   Christy, N. P., and Laragh, J. H. Pathogenesis of hypokalemic alkalosis in Cushing's syndrome. *N. Engl. J. Med.* 265:1083, 1961.

Holmes, A. M., Marrott, P. K., Young, J., and Prentice, E. Pseudohyperaldosteronism induced by habitual ingestion of liquorice. *Postgrad. Med. J.* 46:625, 1970.

McMillan, D. E., and Freeman. R. B. The milk alkali syndrome: A study of the acute disorder with comments on the development of the chronic condition. *Medicine* 44:485, 1965.

van Goidsenhoven, G. M. T., Gray, O. V., Price, A. V., and Sanderson, P. H. The effect of prolonged administration of large doses of sodium bicarbonate in man. *Clin. Sci.* 13:383, 1954.

6 : Respiratory Acidosis: Clinical Examples

Causes The causes of respiratory acidosis are listed in Table 6-1. The difference between acute and chronic respiratory acidosis was described in Chapter 1, under Compensatory Responses (p. 12). The distinction depends on the duration of the acidosis. An *acute* respiratory acidosis becomes manifest within minutes after the introduction of the perturbation. The compensatory rise in serum $[HCO_3^-]$ is slight because it depends entirely on the processing of the retained CO_2 and the buffering of the new H^+, which is produced by this reaction (see Eqs. 1-1, 1-2, and 3-7). Consequently, the decrease in pH tends to be marked in acute respiratory acidosis. In contrast, *chronic* respiratory acidosis develops when the elevated P_{CO_2} is sustained long enough to induce renal compensation, i.e., for at least 3 to 4 days. Often the elevated P_{CO_2} is present for years. The renal compensation consists of a transient increase in the excretion of H^+ (generating new HCO_3^- stores) and a sustained increase in the reabsorption of HCO_3^- (Fig. 2-2B). Consequently, the blood pH is usually not so low in chronic as in acute respiratory acidosis (Eq. 3-6 and Table 3-2).

Severe Pneumonia A 73-year-old man with a lifelong history of smoking cigarettes developed cough, fever, and shortness of breath. Over the next several days he tried to manage his illness at home by increasing his intake of fluids and taking aspirin and cough syrup. However, the cough worsened and it became productive of yellow sputum. The fever continued and the patient became confused, so that he was admitted to the hospital.

On admission the patient's temperature was 38.5°C, the pulse rate was 110 per minute and regular, and the respirations were 28 per minute and labored. The patient was somnolent, and when aroused he evidenced disorientation as to time and place. His lips and extremities appeared cyanotic. The mucous membranes were dry, and the neck veins were flat when the patient lay supine. Examination of the chest revealed dullness to percussion over both lung fields posteriorly; diffuse rales were heard throughout. Cardiac examination showed a regular rhythm without gallop or murmurs. The abdomen was scaphoid and

Table 6-1 : Causes of respiratory acidosis
The causes have been subdivided into those that primarily induce acute respiratory acidosis (up to 8 hours in duration) and those that usually lead to chronic respiratory acidosis (present for 3 days or longer). The distinction between acute and chronic is described in the text.

Acute
  Airway obstruction
    Aspiration
    Laryngospasm
    Severe bronchospasm
  Circulatory catastrophes
    Cardiac arrest
    Severe pulmonary edema
    Massive pulmonary embolism
  Central nervous system depression
    Sedative overdosage
    Cerebral trauma or infarct
  Severe neuromuscular impairment
    High cervical cordotomy
    Severe peripheral nerve disease
    Myasthenic crisis
    Hypokalemic myopathy
  Severe ventilatory restriction
    Severe pneumonia
    Pneumothorax or hemothorax
    Flail chest from trauma
Chronic
  Chronic obstructive pulmonary disease
  Neuromuscular diseases
    Poliomyelitis
    Muscular dystrophy
    Amyotrophic lateral sclerosis
  Ventilatory restriction
    Kyphoscoliosis
    Interstitial fibrosis
    Severe obesity
  Central nervous system disorders
    Primary alveolar hypoventilation
    Obesity-hypoventilation syndrome

no abnormal masses were felt. There was no edema of the extremities.

Chest x-ray showed dense pneumonic infiltrates on both sides. The results of selected laboratory tests are shown in Table 6-2.

Comment Although the venous values are normal, the arterial acid-base composition clearly demonstrates the presence of a rather severe acidosis. The increase in P_{CO_2} to 70 mm Hg, combined with a $[HCO_3^-]$ at the upper limit of normal, indicates that the acidosis is respiratory in origin (Table 3-2). In order to evaluate whether the serum $[HCO_3]$ is in the appropriate range for this degree of hypercapnia, one can use the rules of thumb (Table 3-3) or the confidence bands (Fig. 3-1). Recall that the expected

**Table 6-2 : Results of laboratory tests
in a 73-year-old man with severe pneumonia**
The numbers in parentheses indicate estimated values.

Test	On admission	24 Hours after admission
Arterial blood:		
pH	7.21	7.37
P_{CO_2} (mm Hg)	70	45
$[H^+]$ (nmoles/L)	62 (63)	43 (43)
$[HCO_3^-]$ (mmoles/L)	27 (27)	25 (25)
P_{O_2} (mm Hg)	40	70
Venous serum:		
$[Na^+]$ (mmoles/L)	142	140
$[K^+]$ (mmoles/L)	4.5	4.0
$[Cl^-]$ (mmoles/L)	103	103
Total CO_2 (mmoles/L)	28	26
Anion gap (meq/L)	11	11

rise in serum $[HCO_3^-]$ differs in acute vs. chronic respiratory acidosis (see the beginning of this chapter). Although the present illness in this patient spanned several days, it is likely that the respiratory deficiency that led to the marked retention of CO_2 was more recent in origin. As a first step, therefore, one might think of acute respiratory acidosis in which the $[HCO_3^-]$ rises by approximately 0.1 mmole per liter for each mm Hg rise in P_{CO_2}. In the present instance, where the P_{CO_2} has risen by 30 mm Hg above the normal value, the rules of thumb would predict a rise in $[HCO_3^-]$ of 3 mmoles per liter, for a new $[HCO_3^-]$ of 24 + 3, or 27 mmoles per liter. The fact that this prediction coincides with the measured value (Table 6-2) confirms the diagnostic impression of an uncomplicated acute respiratory acidosis.

Note that the change in $[HCO_3^-]$ is small in acute respiratory disturbances (Fig. 3-1 and Table 3-3); in fact, in the present patient the serum $[HCO_3^-]$ did not rise out of the normal range (Table 1-1). As noted at the beginning of this chapter, the rise is small in acute disturbances because it depends entirely on chemical processing of the retained CO_2 and titration of buffers. Virtually all the increase in $[HCO_3^-]$ is due to the titration of intracellular nonbicarbonate buffers (Eq. 3-7) by the H^+ formed during the hydration or hydroxylation of the retained CO_2 (Eqs. 1-1 and 1-2). No renal generation of new HCO_3^- occurs. During chronic respiratory acidosis there is increased renal reabsorption of new HCO_3^-, and therefore the $[HCO_3^-]$ rises much more than during acute respiratory acidosis (Fig. 3-1B).

As explained in Chapter 3 (under Meaning of Anion Gap, p. 79), the anion gap does not change during uncomplicated respiratory acidosis (Tables 3-4 and 6-2).

The somnolence and disorientation in acute respiratory acidosis are usually ascribed to a direct narcotic effect of the CO_2; in this patient, however, the presence of cyanosis suggests that cerebral hypoxia may have been an important contributing factor. The arterial PO_2 value of 40 mm Hg is consistent with this formulation.

Treatment

Treatment of acute respiratory acidosis should be directed at reducing the PCO_2 promptly. This goal can be accomplished only by inserting an endotracheal tube and assisting the patient's ventilatory efforts by use of a mechanical ventilator. Intubation is mandatory for survival of patients with pneumonia and acute hypercapnia because of the severe impairment in alveolar reserve. The patient's respiratory effort is maximal, and if fatigue develops and his or her effort declines, the patient will almost certainly develop lethal hypoxia and acidosis.

This therapeutic approach, coupled with appropriate antibiotics, rapidly reversed the hypercapnia and acidosis (as well as the hypoxemia) in this patient, so that after 24 hours the arterial acid-base values were normal or very close to normal. Within 48 hours after admission the patient was alert and oriented, and he no longer required support with the ventilator. Follow-up x-ray of the chest showed improvement. The patient was encouraged to stop smoking, and after ten days in the hospital he was discharged.

Acute Pulmonary Edema

A 57-year-old man who had a history of myocardial infarction and was being treated for congestive heart failure collapsed suddenly at home. An ambulance was called, and the emergency medical personnel arrived to find the patient cyanotic and unresponsive but breathing spontaneously. He had a rapid pulse and his blood pressure was 100/60 mm Hg. Oxygen was administered by face mask, and the patient was rushed to a nearby emergency room. There, an endotracheal tube was inserted immediately and the patient was treated with assisted ventilation and 100% oxygen.

On physical examination in the emergency room, the patient's neck veins were found to be markedly distended even when he was sitting up. He was no longer cyanotic, probably because he was receiving oxygen. There were moist rales throughout both lung fields. The heart sounds were faint and rapid, and an S3 gallop rhythm was present. The abdomen was distended, and an enlarged liver was easily palpable. There was 1+ edema of the extremities.

Tracheal suction produced copious pink, frothy fluid. X-ray of the chest showed an enlarged heart and bilateral pulmonary

**Table 6-3 : Results of laboratory tests
in a 57-year-old man with pulmonary edema**
The numbers in parentheses indicate estimated values.

Test	Before intubation	30 Minutes after intubation	24 Hours after admission
Arterial blood:			
pH	7.05	7.29	7.42
PCO_2 (mm Hg)	112	60	40
[H^+] (nmoles/L)	89 (90)	51 (51)	38 (38)
[HCO_3^-] (mmoles/L)	30 (30)	28 (28)	25 (25)
PO_2 (mm Hg)	45	140	120
Venous serum:			
[Na^+] (mmoles/L)	144		140
[K^+] (mmoles/L)	4.9		4.2
[Cl^-] (mmoles/L)	98		100
Total CO_2 (mmoles/L)	33		27
Anion gap (meq/L)	13		13

opacities consistent with severe pulmonary edema. Table 6-3 shows laboratory values on blood taken during the administration of oxygen but before tracheal intubation.

Comment The history, physical examination, and arterial blood measurements all point to severe acute respiratory acidosis. Evaluation of the [HCO_3^-] by using the appropriate rule of thumb (Table 3-3) indicates that the increment is within the expected range — i.e., a predicted rise of $0.1 \times (112 - 40)$, or 7 mmoles per liter, for a new [HCO_3^-] of 31 mmoles per liter $(24 + 7)$.

Acute pulmonary edema does not invariably lead to respiratory acidosis but may be associated with a variety of acid-base disorders. Impairment in the transfer of oxygen from alveolus to blood develops early, and the resultant hypoxemia leads to hyperventilation, so that respiratory alkalosis may be the predominant derangement of acid-base balance. In addition, ventilation may be stimulated under these circumstances by intrapulmonary receptors (see Chap. 7, under Pulmonary Embolism). If the accumulation of fluid in the alveoli is more extensive, transfer of CO_2 from blood to alveolus will be impaired as well, and then respiratory acidosis develops. Moreover, if hypoxemia is severe enough, or if cardiac output falls to levels at which delivery of oxygen to peripheral tissues is sufficiently reduced, metabolic (lactic) acidosis can occur (see Chap. 4); in that event, the patient will have a mixed disturbance of respiratory and metabolic acidosis (an example is discussed in Chap. 8, under Pulmonary Edema and Shock). In the present patient, however, the normal anion gap (Table 6-3) excludes a metabolic component.

Administration of oxygen by members of the ambulance crew was probably lifesaving for this patient. A combined respiratory and metabolic acidosis might have plunged the pH to a fatal level. As can be seen in Table 6-3, the P_{O_2} of only 45 mm Hg despite administration of 100% oxygen by face mask (meaning that perhaps 60 to 80% oxygen was delivered) indicates that transport of oxygen into the blood was severely curtailed. Inequality of the ratio of alveolar ventilation to blood flow between various parts of the lungs (the so-called $\dot{V}_A/\dot{Q}$ ratio) may also have contributed to the hypoxemia. Even when 100% oxygen was given through the endotracheal tube, the P_{O_2} rose to only 140 mm Hg (Table 6-3). Therefore, as in the last patient discussed in this chapter, intubation of the trachea was almost certainly essential for survival, and this measure was appropriately carried out as soon as the patient arrived in the emergency room.

In addition to correcting the acid-base imbalance, therapy was directed toward reversing the cardiac failure. Furosemide was administered to remove fluid from the lungs, and the peripheral vasodilator nitroprusside was given to reduce cardiac work. Over the following 24 hours, pulmonary gas exchange improved to the point where P_{CO_2} returned to normal (Table 6-3), and where P_{O_2} was 120 mm Hg even when the patient was given only 40% oxygen. Several days later the endotracheal tube was removed. As had been suspected from the beginning, the acute pulmonary edema had been precipitated by another myocardial infarction, from which the patient recovered without further complications.

Chronic Obstructive Pulmonary Disease

A 44-year-old woman had had asthma since childhood, and she had been a heavy cigarette smoker since her teens. Over the past four to five years she gradually developed progressive shortness of breath, and she had experienced episodes of somnolence. In addition, she had developed pretibial and ankle edema, for which she had been treated intermittently with diuretics. Because of worsening shortness of breath and occasional episodes of confusion, she was admitted to the hospital for evaluation. At the time of admission she had been off diuretics for several weeks.

On physical examination the patient was alert and oriented, but she appeared short of breath at rest and she coughed frequently. Her temperature was 37°C, her pulse rate 100 per minute; respirations were 25 per minute and shallow. Her chest was barrel-shaped (increased anteroposterior diameter). Breath sounds were distant, and basilar rales as well as expiratory wheezes were present. The heart sounds were distant; no murmurs or gallops were heard. The liver was felt 4 to 5 cm below the right

Table 6-4 : Results of laboratory tests in a 44-year-old woman with chronic obstructive lung disease
The numbers in parentheses indicate estimated values.

Test	On admission	1 Week after admission
Arterial blood:		
pH	7.34	7.37
P_{CO_2} (mm Hg)	65	56
[H$^+$] (nmoles/L)	46 (46)	43 (43)
[HCO$_3^-$] (mmoles/L)	34 (34)	31 (31)
P_{O_2} (mm Hg)	45	55
Venous serum:		
[Na$^+$] (mmoles/L)	138	137
[K$^+$] (mmoles/L)	4.0	4.0
[Cl$^-$] (mmoles/L)	90	94
Total CO$_2$ (mmoles/L)	37	34
Anion gap (meq/L)	11	9

costal margin, but its total span along the craniocaudal axis was normal, approximately 12 cm. The patient had 1+ pretibial edema.

Pertinent laboratory values on admission are shown in Table 6-4.

Comment

The combination of an acid pH (7.34) and an elevated P_{CO_2} indicates that the primary disturbance in this patient is a respiratory acidosis. Evaluation of the serum [HCO$_3^-$] by using the rules of thumb (Table 3-3) or the confidence bands (Fig. 3-1) shows that the [HCO$_3^-$] has risen by more than would be expected in uncomplicated *acute* respiratory acidosis. That is, with an increase in P_{CO_2} of 25 mm Hg (65 − 40) we would anticipate a rise in [HCO$_3^-$] of approximately 3 mmoles per liter (Table 3-3), to a new value of approximately 27 mmoles per liter (24 + 3); yet the measured value was 34 mmoles per liter (Table 6-4). The latter value is, however, within the range expected in *chronic* respiratory acidosis (0.4 × 25). The anion gap is normal in chronic respiratory acidosis (Table 3-4), and the calculated value in this patient is consistent with this prediction. Thus, the laboratory measurements support the diagnosis suggested by the patient's history and physical examination, namely, chronic respiratory acidosis.

The rise in serum [HCO$_3^-$] in chronic respiratory acidosis is due to the following sequence of events in the kidneys. In response to sustained hypercapnia, renal H$^+$ excretion increases transiently (for 2 to 3 days) and in the process new HCO$_3^-$ is generated (Figs. 2-3 and 2-6). This new HCO$_3^-$ is retained in the body by an increase in HCO_3^- reabsorption that develops fully over 3 to 4 days and is sustained for as long as the PCO_2 remains elevated. The additional HCO_3^- mitigates the acidemia (Eq. 1-9), although it does not return the arterial pH completely to normal. Thus, chronic respiratory acidosis is characteristically accompanied by a mild acidemia (Fig. 3-1A). Although pulmonary insufficiency had been present in this patient for many years, the chronic steady state of respiratory acidosis can be reached in a very few days — just as soon as the renal response of increased HCO_3^- reabsorption has developed fully.

As discussed in Chapter 3 (under Meaning of Anion Gap), the increase in the renal reabsorption of HCO_3^- is coupled with a decrease in the reabsorption of Cl^-, so that the serum $[Cl^-]$ is abnormally low. The concurrence of increased serum $[HCO_3^-]$ and decreased $[Cl^-]$, which is characteristic of chronic respiratory acidosis, occurs as well during metabolic alkalosis (e.g., Tables 5-2, 5-3, and 5-5). Consequently, an arterial blood pH (and PCO_2) measurement is required to distinguish the disorders definitively — although in most instances, as in the present patient, the distinction can be made securely on the basis of the history. The serum $[K^+]$ often supplies a helpful clue; it is virtually always low in metabolic alkalosis, whereas it is usually normal in respiratory acidosis.

Although patients with pulmonary insufficiency may have uncomplicated chronic respiratory acidosis, they frequently develop mixed acid-base disorders. The most common of these mixed disorders are acute respiratory acidosis (due, for example, to an intercurrent infection) superimposed on chronic respiratory acidosis, and combined metabolic alkalosis (caused by diuretic therapy) along with chronic respiratory acidosis. These mixed disorders are discussed in Chapter 8 (see also Chap. 3, Confidence Bands, patient 4, p. 72).

Treatment

In contrast to the treatment of acute respiratory acidosis, which is a matter of emergency, that of chronic respiratory acidosis is (and should be) a more gradual process. Although the goal is the same — namely, to reduce the PCO_2 — an abrupt reduction of PCO_2 in chronic respiratory acidosis will lead to severe alkalosis (Eq. 3-6) because the serum $[HCO_3^-]$ will not be reduced so quickly. Therefore, the therapeutic aim should be for a gradual reduction of the PCO_2, as by treating any infection and removing excess fluid with diuretics. In the present patient, this approach led to improvement over the following week, and the PCO_2 eventually stabilized in the mid-50s (Table 6-4), resulting in a nearly normal acid-base and electrolyte status.

Selected
References

General

Brackett, N. C., Jr., Cohen, J. J., and Schwartz, W. B. Carbon dioxide titration curve of normal man: Effect of increasing degrees of acute hypercapnia on acid-base equilibrium. *N. Engl. J. Med.* 272:6, 1965.

Brackett, N. C., Jr., Wingo, C. F., Muren, O., and Solano, J. T. Acid-base response to chronic hypercapnia in man. *N. Engl. J. Med.* 280:124, 1969.

Cogan, M. G. Chronic hypercapnia stimulates proximal bicarbonate reabsorption in the rat. *J. Clin. Invest.* 74:1942, 1984.

Cogan, M. G. Effects of acute alterations in Pco_2 on proximal HCO_3^-, Cl^- and H_2O reabsorption. *Am. J. Physiol.* 246 (Renal Fluid Electrolyte Physiol. 15):F21, 1984.

Cohen, J. J., and Kassirer, J. P. *Acid-Base.* Boston: Little, Brown, 1982. Chap. 10.

Cohen, J. J., and Schwartz, W. B. Evaluation of acid-base balance equilibrium in pulmonary insufficiency: An approach to a diagnostic dilemma. *Am. J. Med.* 41:163, 1966.

Gennari, F. J. Respiratory acidosis and alkalosis. In M. H. Maxwell, C. R. Kleeman, and R. G. Narins (eds.), *Clinical Disorders of Fluid and Electrolyte Metabolism* (3rd ed.). New York: McGraw-Hill, 1986.

Goldring, R. M., Turino, G. M., and Heinemann, H. O. Respiratory-renal adjustments in chronic hypercapnia in man. Extracellular bicarbonate concentration and the regulation of ventilation. *Am. J. Med.* 51:772, 1971.

Goldstein, M. B., Gennari, F. J., and Schwartz, W. B. The influence of graded degrees of chronic hypercapnia on the acute carbon dioxide titration curve. *J. Clin. Invest.* 50:208, 1971.

Manfredi, F. Effects of hypocapnia and hypercapnia on intracellular acid-base equilibrium in man. *J. Lab. Clin. Med.* 69:304, 1967.

Nattie, E. E., and Romer, L. CSF HCO_3^- regulation in isosmotic conditions: The role of brain Pco_2 and plasma HCO_3^-. *Respir. Physiol.* 33:17, 1978.

Schwartz, W. B., Brackett, N. C., Jr., and Cohen, J. J. The response of extracellular hydrogen ion concentration to graded degrees of chronic hypercapnia: The physiologic limits of the defense of pH. *J. Clin. Invest.* 44:291, 1965.

Schwartz, W. B., and Cohen, J. J. The nature of the renal response to chronic disorders of acid-base equilibrium. *Am. J. Med.* 64:417, 1978.

Schwartz, W. B., Hays, R. M., Polak, A., and Haynie, G. D. Effects of chronic hypercapnia on electrolyte and acid-base equilibrium: II. Recovery, with special reference to the influence of chloride intake. *J. Clin. Invest.* 40:1238, 1961.

West, J. B. Causes of carbon dioxide retention in lung disease. *N. Engl. J. Med.* 284:1232, 1971.

Severe Pneumonia

Simpson, H., and Flenley, D. C. Arterial blood-gas tensions and pH in acute lower-respiratory-tract infections in infancy and childhood. *Lancet* 1:7, 1967.

142

Pulmonary Edema   Aberman, A., and Fulop, M. The metabolic and respiratory acidosis of acute pulmonary edema. *Ann. Intern. Med.* 76:173, 1972.

Anthonisen, N. R., and Smith, H. J. Respiratory acidosis as a consequence of pulmonary edema. *Ann. Intern. Med.* 62:991, 1965.

Avery, W. G., Samet, P., and Sackner, M. A. The acidosis of pulmonary edema. *Am. J. Med.* 48:320, 1970.

Chronic Obstructive Lung Disease   Burrows, B., and Earle, R. H. Course and prognosis of chronic obstructive lung disease. *N. Engl. J. Med.* 280:397, 1969.

Lourenco, R. V., and Miranda, J. M. Drive and performance of the ventilatory apparatus in chronic obstructive lung disease. *N. Engl. J. Med.* 279:53, 1968.

Neff, T. A., and Petty, T. L. Tolerance and survival in severe chronic hypercapnia. *Arch. Int. Med.* 129:591, 1972.

Robin, E. D. Abnormalities of acid-base regulation in chronic pulmonary disease, with special reference to hypercapnia and extracellular alkalosis. *N. Engl. J. Med.* 268:917, 1963.

Other Causes   Hopewell, P. C., and Murray, J. F. The adult respiratory distress syndrome. *Ann. Rev. Med.* 27:343, 1976.

Lambertsen, C. J. Effects of drugs and hormones on the respiratory response to carbon dioxide. In W. O. Fenn and H. Rahn (eds.), *Handbook of Physiology,* Section 3, Respiration, vol. 1, Chap. 22. Baltimore: Waverley, 1964.

McFadden, E. R., Jr., and Lyons, H. A. Arterial blood gas tensions in asthma. *N. Engl. J. Med.* 278:1027, 1968.

Mellins, R. B., Balfour, H. H., Turino, G. M., and Winters, R. W. Failure of automatic control of ventilation (Ondine's curse). *Medicine* 49:487, 1970.

Menitove, S. M., and Goldring, R. M. Combined ventilator and bicarbonate strategy in the management of status asthmaticus. *Am. J. Med.* 74:898, 1983.

Miller, A., and Granada, M. In-hospital mortality in the Pickwickian syndrome. *Am. J. Med.* 56:144, 1974.

7 : Respiratory Alkalosis: Clinical Examples

Causes

The causes of respiratory alkalosis are listed in Table 7-1. They are categorized by various stimuli to ventilation, including hypoxemia, drugs and hormones, and diseases of the central nervous system and lungs. The pulmonary diseases that cause respiratory alkalosis can also cause respiratory acidosis (see Table 6-1). Which disorder is induced will depend on the severity of the pulmonary disease. With milder disturbances, hyperventilation is the predominant change, and P_{CO_2} falls; with more severe disease, gas exchange is so impaired that P_{CO_2} rises despite hyperpnea. In a similar fashion, injury to the central nervous system can produce either acid-base disorder, depending on the area of the brain affected and the severity of the damage.

Table 7-1 shows the causes further subdivided into those that primarily cause acute respiratory alkalosis and those that are more likely to produce sustained hyperventilation and chronic respiratory alkalosis. The distinction between acute and chronic respiratory alkalosis is discussed in Chapter 1 and is analogous (although opposite in direction) to that described for respiratory acidosis at the beginning of Chapter 6.

Gram-Negative Sepsis

A 58-year-old woman with a history of recurrent bouts of pyelonephritis was known to have staghorn calculi in both renal pelves, as well as mild renal insufficiency (serum creatinine of 2.0 mg per 100 ml, or 177 μmoles per liter). The day before seeing her physician she developed fever, chills, and generalized aching. She went to bed, took some acetaminophen tablets, and drank extra fluids. The next afternoon, when her fever spiked to 40.5°C, she became confused and lethargic. She was admitted to the hospital.

Physical examination on admission showed the patient to be oriented as to person, but not as to time and place. Her temperature was 39.5°C, respiratory rate 24 per minute and somewhat deep but not labored, and pulse rate 120 per minute and regular. Her skin was clear and warm to touch, and there was no evidence of peripheral vasoconstriction despite a blood pressure of 90/60 mm Hg and the rapid pulse. Her neck was supple. There

Table 7-1 : Causes of respiratory alkalosis
The causes are categorized in two ways: by stimuli to respiration, and by whether the disturbances they produce are primarily acute (up to 8 hours' duration) or chronic (present for 3 days or longer). The conditions listed under other causes are probably mediated by stimulation of the ventilatory center in the brain, although this mechanism has not been conclusively identified.

	Acute	Chronic
Hypoxia		
High altitude		X
Inequality of ventilation-perfusion		X
Hypotension	X	
Central nervous system		
Voluntary hyperventilation	X	
Anxiety-hyperventilation syndrome	X	
Neurologic diseases		
Cerebrovascular accident		X
Infection	X	
Trauma	X	
Tumor		X
Drugs or hormones		
Salicylates	X	
Nicotine	X	
Dinitrophenol	X	
Xanthines		X
Pressor hormones	X	
Progesterone		X
Pulmonary diseases		
Interstitial fibrosis		X
Pneumonia	X	
Pulmonary edema	X	
Pulmonary embolism	X	
Other causes		
Pregnancy		X
Hepatic failure		X
Gram-negative septicemia	X	
Exposure to heat	X	
Mechanical overventilation		X

was no evidence of upper respiratory infection; the lungs were clear on auscultation. Examination of the heart revealed no murmurs, rubs, or gallop. There was no edema. The abdomen was soft, no masses were felt, and there was no evidence of hepatic or splenic enlargement; bowel sounds were hypoactive but present. Pelvic and rectal examination was normal.

Laboratory evaluation included the following results. Chest x-ray was normal. White blood cell count was 12,000/mm^3, with 85% polymorphonuclear leukocytes and 10% band forms. The urinary sediment revealed white blood cells (WBC) too numerous to count, and many WBC clumps as well as occasional WBC casts; there were many bacteria, and a gram stain on a drop of uncentrifuged urine showed gram-negative rods. Blood cultures obtained at the time of admission subsequently grew out gram-negative rods in all bottles. The results of acid-base studies are shown in Table 7-2.

**Table 7-2 : Results of laboratory tests
in a 58-year-old woman with bacterial sepsis**
The numbers in parentheses indicate estimated values.

Test	On admission	48 Hours after admission
Arterial blood:		
pH	7.56	7.44
P_{CO_2} (mm Hg)	22	35
$[H^+]$ (nmoles/L)	28 (28)	36 (36)
$[HCO_3^-]$ (mmoles/L)	19 (19)	23 (23)
P_{O_2} (mm Hg)	110	95
Venous serum:		
$[Na^+]$ (mmoles/L)	138	140
$[K^+]$ (mmoles/L)	3.8	4.0
$[Cl^-]$ (mmoles/L)	105	104
Total CO_2 (mmoles/L)	20	25
Anion gap (meq/L)	13	11

Comment

The arterial acid-base values in this patient indicate the presence of a respiratory alkalosis (Table 3-2): P_{CO_2} is reduced and the pH is notably alkalemic (>7.45). In order to evaluate whether the change in $[HCO_3^-]$ is consistent with uncomplicated respiratory alkalosis, one can use either the rules of thumb (Table 3-3) or the confidence bands (Fig. 3-1). In this instance, the history points to an acute disturbance, where the expected fall in $[HCO_3^-]$ is 0.2 mmole per liter for each mm Hg drop in P_{CO_2} (Table 3-3), or a new, anticipated $[HCO_3^-]$ of approximately 20 mmoles per liter $[0.2 \times (40 - 22) = 3.6; 24 - 4 = 20$ mmoles/L]. The fact that the measured value was 19 mmoles per liter (Table 7-2) is thus consistent with uncomplicated acute respiratory alkalosis. It is often difficult or impossible, however, to distinguish acute from chronic respiratory alkalosis by using either the rules of thumb or the confidence bands. As shown in Figure 3-1, the bands overlap widely, and hence the rules of thumb, which are derived from the bands, give only minor differences except when the reduction in P_{CO_2} is very severe.

The reduction in $[HCO_3^-]$ in acute respiratory alkalosis, although often slight, as in this patient, nonetheless is sufficient to prevent life-threatening alkalemia. Ordinarily, a fall in serum $[HCO_3^-]$ would be accompanied by a reciprocal rise in serum $[Cl^-]$ (see Chap. 3, under Meaning of Anion Gap, p. 79). In acute respiratory alkalosis, however, serum $[Cl^-]$ does not change notably (Table 3-4). Instead, serum $[Na^+]$ falls slightly, due to a shift of Na^+ into cells. As the reaction in Equation 3-7 shifts to the left — i.e., as H^+ is released from nonbicarbonate buffers (Buf^-) within cells — H^+ exits the cell in a process linked to entry of Na^+ (see Fig. 1-7). As a result, the anion gap is characteristically normal in acute respiratory alkalosis even though $[HCO_3^-]$ is reduced (Tables 3-4 and 7-2).

Note that serum $[K^+]$ did not fall markedly (Table 7-2), despite a major increase in pH. This fact reflects a characteristic difference between respiratory and metabolic acid-base disturbances: shifts of K^+ across cell membranes are much smaller in the former than in the latter. The reasons for this difference are not fully understood, but they may involve the following mechanisms. Experimental evidence indicates that shifts of K^+ across cell membranes are influenced by changes in both pH and $[HCO_3^-]$, and that the two influences are somewhat independent of each other. An increase in either pH or $[HCO_3^-]$ favors K^+ entry; a decrease in either favors K^+ exit from cells. Thus, in respiratory alkalosis two countervailing influences are present: The increase in pH favors K^+ entry, and the fall in $[HCO_3^-]$ favors K^+ exit. The net result is little change in serum $[K^+]$.

Thus, the venous serum electrolytes in this patient (Table 7-2) illustrate a useful clinical dictum: If the only abnormality is a slight reduction in the total CO_2, one should think of acute respiratory alkalosis. Often hyperventilation is not a striking feature on physical examination, and the only clue to respiratory alkalosis may be the low total CO_2 in the venous electrolyte profile.

The presence of respiratory alkalosis usually does not pose an emergent situation requiring treatment by itself, but serves, rather, to confirm the presence of, or call attention to, an underlying disorder. The present patient is a case in point. Her long history of recurrent pyelonephritis, coupled with an acute and severe febrile episode, pointed to gram-negative sepsis as a likely diagnosis. Since respiratory alkalosis is seen in approximately 50 percent of such patients (Table 7-1), the presence of this acid-base disturbance lent further support to the initial diagnostic impression, and the patient was immediately treated with antibiotics known to be effective against gram-negative organisms. The infection was rapidly brought under control. Two days later the Pco_2 and pH were almost in the normal range (Table 7-2). The patient became afebrile and alert, and after ten days in the hospital she was discharged on suppressive antibiotic therapy, having also been scheduled for an operation to remove the staghorn calculi. The mechanism by which systemic infection with gram-negative bacteria causes respiratory alkalosis is unclear, but it is thought to be related to a toxic substance elaborated by these organisms. By itself, fever does not produce hypocapnia.

Pulmonary Embolism

A 45-year-old man who was taking 30 mg of prednisone per day for vasculitis suddenly developed pleuritic pain on the right side. The pain persisted, and it was associated with shortness of breath and, eventually, with a small amount of blood-tinged sputum. The patient had had slight ankle edema, as well as a resolving skin rash from the vasculitis, but there had been no symptoms of phlebitis in the legs.

On arrival in the hospital emergency room he was alert and oriented. He complained of pain in the right chest, especially associated with respiratory excursions. He was afebrile, his pulse rate was 110 per minute and regular, and his respirations were 25 per minute and shallow, with splinting on the right side. On auscultation of the lungs there was a localized area of rales on the right side posteriorly, as well as a pleural friction rub in the same area. Examination of the heart revealed tachycardia and slight accentuation of the second sound over the pulmonic area. The abdomen was normal. A slight degree of edema (1+) was present in both legs, but the calves were equal in size and neither tender nor unduly warm.

X-ray of the chest showed a wedge-shaped infiltrate in the right lower lobe posteriorly. The results of blood gas analysis and venous serum electrolyte concentrations are shown in Table 7-3.

Comment

The arterial values indicate the presence of a respiratory alkalosis (Table 3-2). Using the rules of thumb (Table 3-3): After a change in P_{CO_2} of $40 - 27 = 13$ mm Hg, the expected change in $[HCO_3^-]$ would be 13×0.2, or approximately 3 mmoles per liter, for a new $[HCO_3^-]$ of $24 - 3$, or 21 mmoles per liter. This slight reduction of $[HCO_3^-]$ is consistent with this patient's disturbance being an acute process, which is also suggested by the history. As in the previous patient, the only abnormality in the venous serum electrolytes was a minor reduction in the total CO_2 content, a feature, as pointed out, that is characteristic of acute respiratory alkalosis.

Table 7-3 : Results of laboratory tests
in a 45-year-old man with right-sided chest pain
The numbers in parentheses indicate estimated values.

Test	On admission	3 Days later
Arterial blood:		
pH	7.51	7.40
P_{CO_2} (mm Hg)	27	40
$[H^+]$ (nmoles/L)	31 (31)	40 (40)
$[HCO_3^-]$ (mmoles/L)	21 (21)	24 (24)
P_{O_2} (mm Hg)	72	90
Venous serum:		
$[Na^+]$ (mmoles/L)	139	142
$[K^+]$ (mmoles/L)	3.8	4.0
$[Cl^-]$ (mmoles/L)	104	104
Total CO_2 (mmoles/L)	22	26
Anion gap (meq/L)	13	12

In searching for a cause, it is easy in this patient to focus on a disorder of the lungs, and the history is most compatible with a pulmonary embolism. Although the arterial P_{O_2} was reduced below normal (Table 7-3), the reduction is not sufficient to account for the hypocapnia. In order to stimulate ventilation and lower the P_{CO_2}, the P_{O_2} must drop below 60 mm Hg. In this instance, the hyperventilation can be ascribed to stimulation of receptors in the lungs, which send afferent neural signals to the brain. Because of the presence of these receptors, which can respond to injury, hypocapnia is characteristic of pulmonary embolism; this feature is virtually always present, even when the P_{O_2} is only slightly depressed.

All the findings in this patient were so strongly suggestive of pulmonary embolism that he was immediately given heparin for anticoagulation. Subsequently obtained perfusion and ventilation scans of the lungs confirmed the diagnosis. Venous angiograms of the legs showed deep venous thrombosis, presumably a complication of the prednisone therapy. This medication was therefore tapered and then was stopped. Anticoagulation was continued, the lesion in the lungs cleared, and all laboratory values had returned to normal within three days (Table 7-3). The patient was discharged on the oral anticoagulant warfarin. He has had no flare-ups of the vasculitis and remains well.

Anxiety-Hyperventilation Syndrome

A 24-year-old woman was brought to the emergency room by a fellow worker, who reported that the patient had said, "I'm going to die." The day before the patient had had a fight with her boyfriend. She had been up all night and had gone to work without breakfast. At the office she had been distraught, and shortly after arriving there she became light-headed and began to hyperventilate uncontrollably. Soon thereafter, her face felt taut and she began to experience tingling and numbness of the hands and feet. She remembered thinking that she must have had a stroke and that she was going to die. She had taken no drugs.

On arrival in the emergency room she still had the same symptoms. She was conscious, breathing rapidly and deeply, and showed carpopedal spasm. Blood gas analysis on arterial blood taken at this time is shown in Table 7-4. It is of interest that the patient's father had died of a cerebrovascular accident two years earlier.

Comment

Although the presentation of the anxiety-hyperventilation syndrome is usually very characteristic, as it was in this patient, it is nevertheless wise to obtain a blood gas analysis before treatment is instituted. Since metabolic acidosis or pulmonary embo-

Table 7-4 : Results of blood gas analysis in a 24-year-old woman with anxiety-hyperventilation syndrome
The numbers in parentheses indicate estimated values.

Test	On arrival in emergency room	10 Minutes Later
Arterial blood:		
pH	7.56	7.41
P_{CO_2} (mm Hg)	23	41
[H^+] (nmoles/L)	28 (28)	39 (39)
[HCO_3^-] (mmoles/L)	20 (20)	25 (25)
P_{O_2} (mm Hg)	95	90

lism can give a similar clinical picture, it is important to know the initial acid-base values in the event a therapeutic trial (see below) does not immediately reverse the picture. In the present instance the physician felt so sure of the diagnosis that she began treatment as soon as the history and physical examination had been completed and the arterial sample taken. The patient was reassured, and was asked to breathe in and out from a paper bag. Within a few minutes she calmed down and her signs and symptoms disappeared. Analysis of arterial blood 10 minutes after arrival at the hospital showed a return to normal acid-base status (Table 7-4). The physician spoke with the patient for another 45 minutes, during which period she explained the dynamics of the anxiety-hyperventilation syndrome. The patient returned to work later that morning.

In retrospect, the blood gas analysis on arrival (Table 7-4) was consonant with the diagnosis. The P_{CO_2} was reduced by 17 mm Hg (from 40 to 23), and in an acute respiratory alkalosis (Table 3-3) that change would be expected to lower the [HCO_3^-] by approximately 3 mmoles per liter (17 $\times$ 0.2), for a new [HCO_3^-] of approximately 21 mmoles per liter (24 − 3).

Two factors may have contributed to the patient's light-headedness. She may have had hypoglycemia, since she had had nothing to eat. In addition, an acute decrease in P_{CO_2} and the resultant alkalemia reduce cerebral blood flow — sometimes by as much as 40% — leading not only to dizziness, but also to impairment of cerebral function. The patient's other neurological manifestations (tingling, numbness, and carpopedal spasm) were also due to the alkalemia, and in part to the decrease in the plasma concentration of ionized calcium that results from alkalemia.

Both the onset and cessation of acute respiratory disturbances can be very rapid. Arterial pH begins to rise within 15 to 20 seconds after the beginning of hyperventilation, and for any degree of hyperventilation a new steady-state value develops within 10 to 15 minutes. Similarly, within a few minutes after the patient began to inspire air with a high Pco_2 (having filled the paper bag with CO_2 from her own expiration), her arterial pH decreased, as reflected in the disappearance of the neurological symptoms and signs.

Hepatic Failure   A 58-year-old man with a long history of heavy alcohol ingestion had been admitted to the hospital numerous times over the last five years because of jaundice and ascites. On one of these admissions cirrhosis had been diagnosed on the basis of a liver biopsy. The present admission, like those during the prior nine months, was precipitated by the development of confusion and lethargy, rather than by a flare-up of jaundice and ascites.

On entry into the hospital the patient was found to be icteric, and his skin showed numerous spider angiomata. He was oriented as to person but not as to time or place, and he had asterixis. His temperature was normal, his blood pressure 105/60 mm Hg, his pulse rate 80 per minute and regular, and his respiratory rate 20 per minute but not labored. The lungs and heart were normal. On examination of the abdomen there was moderate ascites as well as an enlarged spleen, and a small, very firm liver could be palpated in the right upper quadrant. There was a trace of edema in both lower extremities. Table 7-5 gives the results of laboratory tests performed at the time of admission.

Comment   Hepatic disease can be associated with a variety of acid-base disorders. Besides respiratory alkalosis (Table 7-1), patients such as the present one may have renal tubular acidosis (RTA), which will lead to changes in venous total CO_2 and $[Cl^-]$ identical to those shown in Table 7-5. A reduction in total CO_2 and an eleva-

**Table 7-5 : Results of laboratory tests
in a 58-year-old man with hepatic failure**
The numbers in parentheses indicate estimated values.

Test	On admission	5 Days later
Arterial blood:		
pH	7.47	7.46
Pco_2 (mm Hg)	20	22
$[H^+]$ (nmoles/L)	34 (34)	35 (34)
$[HCO_3^-]$ (mmoles/L)	14 (14)	15 (16)
Po_2 (mm Hg)	85	82
Venous serum:		
$[Na^+]$ (mmoles/L)	138	142
$[K^+]$ (mmoles/L)	3.6	3.7
$[Cl^-]$ (mmoles/L)	108	112
Total CO_2 (mmoles/L)	15	16
Anion gap (meq/L)	15	14

tion of $[Cl^-]$ are equally compatible with respiratory alkalosis or metabolic acidosis. Patients with cirrhosis may also develop a metabolic alkalosis, if vomiting is a feature of their illness or if they are treated aggressively with diuretics. It is not surprising, therefore, that such patients often have mixed disturbances of H^+ balance (see Chap. 8). Consequently, an arterial blood gas analysis is particularly necessary in patients with cirrhosis in order to clarify their acid-base status. This analysis in the present patient (Table 7-5) clearly revealed a respiratory alkalosis. The history and laboratory values suggested a chronic process. By the rules of thumb (Table 3-3), the expected fall in $[HCO_3^-]$ would be approximately 8 mmoles per liter ($40 - 20 = 20$ mm Hg drop in Pco_2; $20 \times 0.4 = 8$), for a new $[HCO_3^-]$ of 16 mmoles per liter. Although this computation supports the diagnosis of chronic respiratory alkalosis, the confidence bands for respiratory alkalosis (Fig. 3-1) overlap to such an extent that differentiation between acute and chronic forms is difficult based solely on this tool (see above, p. 145).

For unknown reasons, chronic respiratory alkalosis is characteristically accompanied by a very slight rise in the anion gap (see Table 3-4, and last paragraph under Meaning of Anion Gap, Chap. 3). The rise is often so small (approximately 3 meq/L) that the anion gap is merely at the upper limit of normal, as in the present patient (Table 7-5). The rise can be observed, however, if one obtains repeated serum samples during the transition from the acute to the chronic phase of respiratory alkalosis (see Table 7-6 and the discussion of the next patient).

Perusal of Table 7-1 points to hepatic failure as the most likely cause of the sustained hypocapnia in this patient. He did not manifest hypoxia, he had not taken drugs that stimulate ventilation, and he did not have pulmonary disease; nor did he present clear evidence of a primary disorder of the central nervous system. The factor stimulating ventilation in hepatic failure is not known, although ammonia has been implicated.

As can be seen from the follow-up laboratory results, there was little improvement in the Pco_2 despite optimal therapy for the liver disease while the patient was hospitalized. Sustained respiratory alkalosis is a grave prognostic sign in liver failure (survival rarely exceeds six months once this disorder develops). Thus it was in this patient, who died three weeks later in this, his final hospital admission.

Cerebrovascular Accident

A 78-year-old woman was admitted to the hospital in coma after collapsing at home. She had a history of hypertension, and had had several "small strokes," although she had been alert up until

Table 7-6 : Results of laboratory tests in a comatose 78-year-old woman
The numbers in parentheses indicate estimated values.

Test	On admission	1 Week later
Arterial blood:		
pH	7.53	7.51
P_{CO_2} (mm Hg)	25	17
$[H^+]$ (nmoles/L)	30 (30)	31 (31)
$[HCO_3^-]$ (mmoles/L)	20 (20)	13
P_{O_2} (mm Hg)	95	105
Venous serum:		
$[Na^+]$ (mmoles/L)	139	141
$[K^+]$ (mmoles/L)	3.7	3.9
$[Cl^-]$ (mmoles/L)	105	111
Total CO_2 (mmoles/L)	22	14
Anion gap (meq/L)	12	16

the day of admission. On arrival at the hospital she was unresponsive, but she was breathing spontaneously; respirations were deep and regular at 22 per minute. Blood pressure was 180/80 mm Hg, pulse rate 80 per minute and regular. There were no localizing neurological signs, and the physical examination was otherwise normal.

A CT scan that was done shortly after admission to the neurology service showed an infarct in the region of the pons. The results of laboratory tests done at this time are shown in Table 7-6. One week later she remained comatose. Her vital signs were stable. Respirations continued to be deep and regular.

Comment

This patient presents an example of the transition from an acute to a chronic respiratory alkalosis. As suggested by the history and confirmed by the laboratory values (Table 7-6), she had an acute respiratory alkalosis on admission: For a decrease in P_{CO_2} of 15 mm Hg, the rules of thumb (Table 3-3) predict a drop in $[HCO_3^-]$ of 15 × 0.2, or 3 mmoles per liter, and thus a new $[HCO_3^-]$ of approximately 24 − 3, or 21 mmoles per liter. As explained at the beginning of Chapter 6 and stated in Table 3-2, an acute respiratory disturbance involves the titration of body buffers, as shown in Equation 3-7. As the chronic state develops (approximately 2 to 4 days), there is, in addition, a renal compensation, which in the case of respiratory alkalosis involves a transient decrease in the excretion of H^+ and a sustained decrease in the reabsorption of HCO_3^-. Note in Table 7-6 that the arterial pH became slightly less alkalemic even though the P_{CO_2} dropped further, the reason being that renal compensation simultaneously decreased the $[HCO_3^-]$ (Eq. 1-9). In conformity with chronic respiratory alkalosis (Table 3-3), the $[HCO_3^-]$ decreased by approximately 0.4 mmoles per liter for each mm Hg decline in P_{CO_2} (40 − 17 = 23; 23 × 0.4 = 9.2), yielding a new predicted $[HCO_3^-]$ of approximately 24 − 9, or 15 mmoles per liter.

As explained in Chapter 3 (last two paragraphs under Meaning of Anion Gap), the change in $[HCO_3^-]$ during chronic respiratory disturbances is accompanied by an opposite change in $[Cl^-]$. Thus, in the present patient, while renal reabsorption of HCO_3^- decreased during the week following her admission, causing arterial $[HCO_3^-]$ and venous total CO_2 to drop (Table 7-6), the reabsorption of Cl^- and hence venous $[Cl^-]$ rose. Simultaneously, the serum concentration of Na^+ remained relatively constant. Consequently, the anion gap changed very little despite a marked decrease in $[HCO_3^-]$. As noted in the discussion of the preceding patient, however, for unknown reasons the anion gap does rise slightly during chronic respiratory alkalosis, and this rise was recorded in the present patient (Table 7-6).

Hyperventilation due to disorders of the central nervous system, as in this patient, may assume one of two patterns. With diffuse cortical damage there is Cheyne-Stokes ventilation, a type of periodic breathing caused by a hyperreflexic response of the respiratory center to CO_2. With lesions of the midbrain, a constant regular hyperventilation is observed, as in the present instance. Respiratory alkalosis is a grave prognostic sign when it is precipitated by diseases of the central nervous system; mortality approaches 100 percent when the P_{CO_2} is sustained below 20 mm Hg. The present patient succumbed without regaining consciousness.

High Altitude

A mountain climber, who is in good health, volunteered to participate in a scientific expedition to an altitude of 4,300 m (approximately 14,000 ft). Arterial and venous blood samples were obtained before departure and after two weeks of residence at the high altitude (Table 7-7). The volunteer experienced some shortness of breath, even at rest, when he first arrived at the high altitude, but within four days this symptom disappeared.

Comment

In this instance the drive to hyperventilation is hypoxemia. The ambient partial pressure of oxygen at 4,300 m is approximately 85 mm Hg, as opposed to 150 mm Hg at sea level. As a result, alveolar and hence arterial P_{O_2} fall to levels (Table 7-7) at which ventilation is stimulated by afferent signals from peripheral chemoreceptors. When P_{O_2} falls below 60 mm Hg, hyperventilation and hypocapnia develop as a result of this mechanism. In a person with normal lungs, the degree of hypocapnia is a direct function of the degree of hypoxemia. The present example, with a P_{O_2} of 49 mm Hg, represents only moderate hypoxemia. A P_{O_2}

Table 7-7 : Laboratory results in a mountain climber at sea level, and after 2 weeks' residence at 4,300 m (ca. 14,000 ft)
The numbers in parentheses indicate estimated values.

Test	Sea level	2 Weeks at high altitude
Arterial blood:		
pH	7.41	7.47
P_{CO_2} (mm Hg)	41	23
$[H^+]$ (nmoles/L)	39 (39)	34 (34)
$[HCO_3^-]$ (mmoles/L)	25 (25)	16 (16)
P_{O_2} (mm Hg)	87	49
Venous serum:		
$[Na^+]$ (mmoles/L)	140	140
$[K^+]$ (mmoles/L)	4.0	4.0
$[Cl^-]$ (mmoles/L)	100	107
Total CO_2 (mmoles/L)	27	17
Anion gap (meq/L)	13	16

as low as 28 mm Hg has been recorded at the summit of Mt. Everest (8,848 m, or 29,029 ft), with a resultant P_{CO_2} of <10 mm Hg.

The cause of increased ventilation differed with each of the examples of chronic respiratory alkalosis that we have just considered, namely, hepatic failure, cerebrovascular accident, and high altitude. Nonetheless, the response to sustained hypocapnia was identical. The paired measurements in the mountain climber (Table 7-7) illustrate the typical pattern: a decrease in $[HCO_3^-]$, an increase in serum $[Cl^-]$, essentially no change in serum $[Na^+]$, and a minimal rise in the anion gap. Note that the decrease in $[HCO_3^-]$ as the mountain climber acclimated to high altitude is of the order predicted by the rules of thumb (Table 3-3) for an uncomplicated chronic respiratory alkalosis.

When a person returns to sea level, the acid-base status will return to normal over a period of four to five days. Because the P_{CO_2} rises to normal more rapidly than does the $[HCO_3^-]$ (the latter being a slower renal adjustment), there will be a transient period of acidosis (Eq. 3-6).

There is, finally, an interesting lesson about balance to be gained from this example. In the steady state of chronic respiratory alkalosis and hypocapnia, the individual may still be on a normal diet, as was the mountain climber. That individual will therefore produce the normal daily load of nonvolatile or fixed H^+ ions (see beginning of Chap. 1), which must be excreted by the kidney. Accordingly, the urinary pH will be appropriately acid even though the person has systemic alkalemia.

Studies of the type in which this person volunteered have increased our understanding of the responses to respiratory alkalosis, and they have helped in preparing humans for sojourns at high altitude.

Selected References

General

Arbus, G. S., Hebert, L. A., Levesque, P. R., Etsten, B. E., and Schwartz, W. B. Characterization and clinical application of the "significance band" for acute respiratory alkalosis. *N. Engl. J. Med.* 280:117, 1969.

Cogan, M. G. Effects of acute alterations in Pco_2 on proximal HCO_3^-, Cl^- and H_2O reabsorption. *Am. J. Physiol.* 246 (Renal Fluid Electrolyte Physiol. 15):F21, 1984.

Cohen, J. J., and Kassirer, J. P. *Acid-Base.* Boston: Little, Brown, 1982. Chap. 11.

Cohen, J. J., Madias, N. E., Wolf, C. J., and Schwartz, W. B. Regulation of acid-base equilibrium in chronic hypocapnia. Evidence that the response of the kidney is not geared to the defense of extracellular $[H^+]$. *J. Clin. Invest.* 57:1483, 1976.

Gennari, F. J. Respiratory acidosis and alkalosis. In M. H. Maxwell, C. R. Kleeman, and R. G. Narins (eds.). *Clinical Disorders of Fluid and Electrolyte Metabolism* (3rd ed.). New York: McGraw-Hill, 1986.

Gennari, F. J., Goldstein, M. B., and Schwartz, W. B. The nature of the renal adaptation to chronic hypocapnia. *J. Clin. Invest.* 51:1722, 1972.

Gledhill, N., Beirne, G. J., and Dempsey, J. A. Renal response to short-term hypocapnia in man. *Kidney Int.* 8:376, 1976.

Kazemi, H., Shannon, D. C., and Carvallo-Gil, E. Brain CO_2 buffering capacity in respiratory acidosis and alkalosis. *J. Appl. Physiol.* 22:241, 1967.

Kety, S. S., and Schmidt, C. F. The effects of active and passive hyperventilation on cerebral blood flow, cerebral oxygen consumption, cardiac output, and blood pressure of normal young men. *J. Clin. Invest.* 25:107, 1946.

Manfredi, F. Effects of hypocapnia and hypercapnia on intracellular acid-base equilibrium in man. *J. Lab. Clin. Med.* 69:304, 1967.

Tenny, S. M., and Lamb, T. W. Physiological consequences of hypoventilation and hyperventilation. In W. O. Fenn and H. Rahn (eds.), *Handbook of Physiology,* Section 3, Respiration, vol. 2. Washington, D.C.: American Physiological Society, 1965.

Gram-Negative Sepsis

Blair, E. Acid-base balance in bacteremic shock. *Arch. Intern. Med.* 127:731, 1971.

Simmons, D. H., Nicoloff, J., and Guze, L. B. Hyperventilation and respiratory alkalosis as signs of gram-negative bacteremia. *J. Amer. Med. Assoc.* 174:2196, 1960.

Pulmonary Embolism

Horres, A. D., and Bernthal, T. Localized multiple minute pulmonary embolism and breathing. *J. Appl. Physiol.* 16:842, 1961.

Kornbluth, R. S., and Turino, G. M. Respiratory control in diffuse interstitial lung disease and diseases of the pulmonary vasculature. *Clin. Chest Med.* 1:91, 1980.

Stein, M., and Levy, S. E. Reflux and humoral response to pulmonary embolism. *Progr. Cardiovasc. Dis.* 17:167, 1974.

Szucs, M. M., Brooks, H. L., Grossman, W., Banas, J. S., Meister, S. G., Dexter, L., and Dalen, J. E. Diagnostic sensitivity of laboratory findings in acute pulmonary embolism. *Ann. Intern. Med.* 74:161, 1971.

Anxiety-
Hyperventilation
Syndrome

Magarian, G. J. Hyperventilation syndromes: Infrequently recognized expressions of anxiety and stress. *Medicine* 61:219, 1982.

Rice, R. L. Symptom patterns of the hyperventilation syndrome. *Am. J. Med.* 8:691, 1950.

Soley, M. H., and Shock, N. W. The etiology of effort syndrome. *Am. J. Med. Sci.* 196:840, 1938.

Hepatic Failure

Karetzky, M. S., and Mithoefer, J. C. The cause of hyperventilation and arterial hypoxia in patients with cirrhosis of the liver. *Am. J. Med. Sci.* 254:797, 1967.

Prytz, H., and Thomsen, A. C. Acid-base status in liver cirrhosis. Disturbances in stable, terminal and porta-caval shunted patients. *Scand. J. Gastroenterol.* 11:249, 1976.

Tylor, M. P., and Sieker, H. O. Biochemical blood gas and peripheral circulatory alterations in hepatic coma. *Am. J. Med.* 27:50, 1959.

Vanamee, P., Popell, J. W., Glicksman, A. S., Randall, H. T., and Roberts, K. E. Respiratory alkalosis in hepatic coma. *Arch. Intern. Med.* 97:762, 1956.

Cerebrovascular
Accident

Brown, H. W., and Plum, F. The neurologic basis of Cheyne-Stokes respiration. *Am. J. Med.* 30:849, 1961.

Lane, D. J., Rout, M. W., and Williamson, D. H. Mechanisms of hyperventilation in acute cerebrovascular accidents. *Br. Med. J.* 3:9, 1971.

Plum, F., and Swanson, A. G. Central neurogenic hyperventilation in man. *Arch. Neurol. Psychiatry* 81:535, 1959.

Rout, M. W., Lane, D. J., and Wollner. L. Prognosis in acute cerebrovascular accidents in relation to respiratory pattern and blood gas tensions. *Br. Med. J.* 3:7, 1971.

High Altitude

Dempsey, J. A., Forster, H. V., and Dopico, G. A. Ventilatory acclimatization to moderate hypoxemia in man. *J. Clin. Invest.* 53:1091, 1974.

Dill, D. B., Talbott, J. H., and Consolazio, W. V. Blood as a physicochemical system: XII. Man at high altitudes. *J. Biol. Chem.* 118:649, 1937.

Lenfant, C., and Sullivan, K. Adaptation to high altitude. *N. Engl. J. Med.* 284:1298, 1971.

Valtin, H., Tenney, S. M., and Larson, S. L. Carbon dioxide diuresis at high altitude. *Clin. Sci.* 22:391, 1962.

West, J. B. Human physiology at extreme altitudes on Mount Everest. *Science* 223:784, 1984.

West, J. B., and Lahiri, S. (eds.). *High Altitude and Man.* Bethesda: American Physiological Society, 1984.

Other Causes   Fauber, H. R., Yiengst, M. J., and Shock, N. W. The effect of therapeutic doses of aspirin on the acid-base balance of the blood in normal adults. *Am. J. Med. Sci.* 217:256, 1949.

Gabow, P. A., Anderson, R. J., Potts, D. E., and Schrier, R. W. Acid-base disturbances in the salicylate-intoxicated adult. *Arch. Intern. Med.* 138:1481, 1978.

Lucius, H., Gahlenbeck, H., Kleine, H. O., Fabel, H., and Bartels, H. Respiratory functions, buffer system and electrolyte concentrations of blood during human pregnancy. *Respir. Physiol.* 9:311, 1970.

Lyons, H. A., and Antonio, R. The sensitivity of the respiratory center in pregnancy and after the administration of progesterone. *Trans. Assoc. Am. Physicians* 72:173, 1959.

Miller, L. C., Schilling, A. F., Logan, D. L., and Johnson, R. L. Potential hazards of rapid smoking as a technic for the modification of smoking behavior. *N. Engl. J. Med.* 297:590, 1977.

8 : Mixed Acid-Base Disturbances: Clinical Examples

Types of Mixed Disturbances

Mixed disorders occur when two or more primary acid-base disturbances are present simultaneously. Their identification is accomplished through application of the same rules of thumb and confidence bands that we have cited in the preceding chapters. Some of the more common types of mixed disturbances are listed in Table 8-1.

Overdose with Salicylate

A 14-year-old boy with a history of emotional disturbance had an argument with his mother. Several hours later, when the mother entered his room, she found him to be extremely lethargic and breathing very deeply. There was an empty bottle of aspirin next to him. The boy was rushed to the hospital.

On arrival in the emergency room the patient was very drowsy, although he could be aroused. He was uncommunicative. His pulse and blood pressure were normal, but his breathing was deep, labored, and periodic, at a rate of approximately 30 per minute (Kussmaul respiration). Samples of arterial and venous blood were obtained at once, following which emergency treatment for presumed overdosage with salicylate was started. Table 8-2 shows the results of the tests.

Table 8-1 : Examples of mixed acid-base disorders

Mixed respiratory/metabolic
  Respiratory acidosis
    Acute plus metabolic acidosis
    Chronic plus metabolic alkalosis
    Chronic plus metabolic acidosis
  Respiratory alkalosis
    Acute plus metabolic acidosis
    Acute plus metabolic alkalosis
Mixed respiratory
  Acute plus chronic respiratory acidosis
  Acute plus chronic respiratory alkalosis
Mixed metabolic
  Metabolic alkalosis plus metabolic acidosis
Triple disorders
  Acute plus chronic respiratory acidosis plus metabolic alkalosis
  Acute respiratory alkalosis plus metabolic alkalosis plus metabolic
    acidosis

Table 8-2 : Results of laboratory tests in a 14-year-old boy who ingested an overdose of aspirin
The numbers in parentheses indicate estimated values.

Test	On admission	48 Hours after admission	5 Days later
Arterial blood:			
pH	7.45	7.46	7.40
P_{CO_2}(mm Hg)	15	22	40
[H^+] (nmoles/L)	35 (35)	34 (35)	40 (40)
[HCO_3^-] (mmoles/L)	10 (10)	15 (15)	24 (24)
Venous serum:			
[Na^+] (mmoles/L)	142	140	140
[K^+] (mmoles/L)	4.3	3.9	4.0
[Cl^-] (mmoles/L)	104	114	102
Total CO_2 (mmoles/L)	11	16	26
Anion gap (meq/L)	27	10	12
Salicylate (mg/100 ml)	100	0	

Comment

The combination of an arterial pH above 7.40 and a low P_{CO_2} indicates the presence of a respiratory alkalosis (Table 3-2). It is immediately apparent, however, that the [HCO_3^-] of 10 mmoles per liter is lower than what would be predicted from either the rules of thumb (Table 3-3) or the confidence bands (Fig. 3-1). Even in chronic respiratory alkalosis, where the compensatory decrease in [HCO_3^-] is greater than in the acute form (Table 3-2), a decrease in P_{CO_2} of 25 mm Hg (40 − 15) would be accompanied by a fall in [HCO_3^-] of only 10 mmoles per liter (25 × 0.4; Table 3-3), whereas in the present instance the [HCO_3^-] has declined by 14 mmoles per liter (24 − 10; Table 8-2). This finding raises the suspicion that the [HCO_3^-] has been reduced by some factor in addition to hypocapnia, i.e., by a metabolic acidosis — a conclusion that is borne out by the elevated anion gap (Table 8-2). (Recall [Table 3-4] that the gap is not increased during acute respiratory alkalosis, and that it rises only slightly during chronic respiratory alkalosis.)

The same important conclusion can be reached if one analyzes the data from the point of view of a metabolic acidosis; that is, the combination of a low [HCO_3^-] and an anion gap exceeding 20 meq per liter (Table 8-2) signals the presence of a metabolic acidosis. However, the application of either the rules of thumb (Table 3-3) or the confidence bands (Fig. 3-2) demonstrates that the P_{CO_2} of 15 mm Hg (Table 8-2) is much lower than what is to be expected in uncomplicated metabolic acidosis; and, of course, a pH above 7.40 itself says that this is not simply an acidosis.

Thus, either approach indicates that two disturbances are present: a respiratory alkalosis and a metabolic acidosis. This mixed acid-base disorder occurs commonly in salicylate intoxication. Aspirin acts directly on the brain stem to stimulate ventilation, and ingestion of moderate to large amounts of the drug therefore uniformly produces respiratory alkalosis. At the same time, aspirin affects cell metabolism in many ways. It uncouples oxidative phosphorylation, an effect that increases CO_2 production. This increase is more than compensated for by the increase in ventilation. It also has inhibitory effects on many enzyme systems, one result of which is a marked increase in the production of organic acids. In a given individual, the pH after salicylate intoxication will reflect an interplay between (1) the hypocapnia induced by stimulation of ventilation, and (2) the consumption of HCO_3^- by titration of newly produced organic acids. Most young children with salicylate intoxication are acidemic, whereas patients over the age of 10 years tend to be alkalemic.

Treatment

The presence of two primary acid-base disturbances has major implications for therapy. In this instance, for example, administration of large amounts of HCO_3^- to treat the metabolic acidosis has the potential to produce life-threatening alkalemia because there will be continued stimulation of ventilation by the aspirin. Hence, treatment should be directed first at getting rid of the drug.

The kidneys handle salicylate through a combination of glomerular filtration, tubular secretion, and tubular reabsorption. The first two processes promote the excretion of salicylate from the body, whereas the third tends to retard excretion. Therefore, the rationale of therapy is to minimize tubular reabsorption. That task is accomplished by two means: inducing a diuresis and alkalinizing the urine. Administration of either saline or mannitol will depress the reabsorption of salt and water, which in turn will decrease the reabsorption of salicylate by reducing the concentration of salicylate in the renal tubular fluid. Alkalinizing the urine will retard the reabsorption of salicylate by minimizing nonionic diffusion (see Chap. 2, under Nonionic Diffusion, p. 48). Acetylsalicylic acid is a weak acid, which, like other weak acids (Fig. 2-5), exists in part as the nonionized acid and in part as the ionized conjugate base acetylsalicylate. The latter form, being lipid-insoluble, cannot readily traverse the renal tubular epithelium. By rendering the urine alkaline as compared to blood, one can reduce the concentration of the nonionized acid in the tubular fluid and thereby retard reabsorption of the drug. Fortunately, this goal can be accomplished by giving a small amount of HCO_3^- because the existing acute hypocapnia will retard bicarbonate reabsorption in the proximal tubule (Fig. 2-

2B) and cause urinary pH to rise rapidly in response to administration of alkali.

These maneuvers were carried out in this patient, and the concentration of salicylate in serum fell rapidly to zero over the next 48 hours, a fact reflected also in the return of the anion gap to normal (Table 8-2). Note, however, that hypocapnia and a low [HCO_3^-] persisted. This pattern is typical of the response to treatment for salicylate intoxication. As the salicylates are removed from the body, metabolism of endogenous acids returns to normal. Nevertheless, a component of metabolic acidosis persists because during the above-described treatment a deficit of HCO_3^- has been incurred by the diuresis and alkalinization of the urine, which has led to increased excretion not only of acetylsalicylate but also of the anions of other endogenous organic acids (see Chap. 2, under Replenishment of Depleted HCO_3^- Stores, p. 42). At the same time, salicylate-stimulated hyperventilation and hypocapnia often persist for several days. Until ventilation returns to normal, the kidneys will respond in a predictable fashion to the hypocapnia by decreased reabsorption of HCO_3^- (Fig. 2-2B), and, reciprocally, increased reabsorption of Cl^-. Consequently, the serum [Cl^-] rises, and the electrolyte and acid-base patterns become similar to that seen in chronic respiratory alkalosis (Table 8-2). Over the next week, as ventilation returns to normal, P_{CO_2} rises, allowing renal H^+ secretion to increase and thereby restoring [HCO_3^-] to normal levels.

The patient was discharged, to be followed as an outpatient by members of the psychiatry and social service departments.

Severe Emphysema and Cor Pulmonale Treated with Furosemide

A 58-year-old man with severe emphysema developed progressive peripheral edema, for which he was given furosemide. Although the edema could be controlled with increasing doses of the diuretic, the patient became somnolent. He was therefore admitted to the hospital for evaluation.

On admission he was drowsy but awake and oriented. His respirations were slightly labored, occurring at a rate of approximately 15 per minute. Temperature, blood pressure, and pulse were normal. Examination of the chest showed an increased anteroposterior diameter. Breath sounds were distant, with dry rales heard at the bases. The heart sounds also were distant; there were no murmurs or gallops. The liver was felt 4 cm below the right costal margin, but was normal in span. There was a trace of ankle edema. Results of laboratory tests are presented in Table 8-3.

Table 8-3 : Results of laboratory tests in a 58-year-old man with emphysema, hypercapnia, and somnolence
The numbers in parentheses indicate estimated values.

Test	On admission	1 Week later
Arterial blood:		
pH	7.44	7.39
P_{CO_2} (mm Hg)	65	55
[H^+] (nmoles/L)	36 (36)	41 (41)
[HCO_3^-] (mmoles/L)	43 (43)	32 (32)
P_{O_2} (mm Hg)	45	58
Venous serum:		
[Na^+] (mmoles/L)	137	139
[K^+] (mmoles/L)	3.5	4.2
[Cl^-] (mmoles/L)	78	90
Total CO_2 (mmoles/L)	45	34
Anion gap (meq/L)	14	15

Comment

From the finding of a notably high arterial [HCO_3^-] of 43 mmoles per liter, one would expect a higher pH than 7.44 if this were an uncomplicated metabolic alkalosis. This fact immediately raises the suspicion that an acidosis is present as well, and the patient's history and high P_{CO_2} (Table 8-3) suggest that the additional acid-base disturbance is a chronic respiratory acidosis. The diagnostic impression of a mixed disorder can be confirmed by approaching the analysis using the guidelines (Table 3-3 and Figs. 3-1 and 3-2) for either disorder.

In the case of a metabolic alkalosis, the expected compensatory increase in P_{CO_2} would be approximately $0.7 \times (43 - 24)$, or 13 mm Hg, for a new P_{CO_2} of $40 + 13$, or 53 mm Hg. The observed P_{CO_2} of 65 mm Hg is thus much higher than one would expect in an uncomplicated metabolic alkalosis. Alternatively, if one approaches the analysis from the perspective of respiratory acidosis — in this case, chronic — then the expected compensatory rise in [HCO_3^-] would be $0.4 \times (65 - 40)$, or 10 mmoles per liter, for a new [HCO_3^-] of $24 + 10$, or 34 mmoles per liter. Again, the observed [HCO_3^-] is well above this level. Thus, in either case, the conclusion is that the patient has a mixed acid-base disturbance of metabolic alkalosis and respiratory acidosis. The respiratory acidosis in this patient is due to the emphysema (see Chap. 6, under Chronic Obstructive Pulmonary Disease), and the metabolic alkalosis can be explained by the diuretic therapy (Chap. 5, under Administration of a Diuretic). The latter diagnosis is supported by the slight reduction in serum [K^+].

Identification and correction of metabolic alkalosis is particularly important in patients with hypercapnia, because the alkalemia may inhibit ventilation — despite the presence of hypoxemia (Table 8-3) — and thereby worsen the hypercapnia. It is likely, for example, that the somnolence that led to this patient's hospitalization was due to increasing hypercapnia precipitated by metabolic alkalosis. In the present patient, the dual measures of withholding the diuretic medication and giving KCl sufficed to lower not only the $[HCO_3^-]$ but also the PCO_2 (Table 8-3). The patient's alertness improved, and he was discharged without diuretic medication and urged to control his edema through restriction of salt in his food.

Pulmonary Edema and Shock

A 74-year-old woman who was known to have arteriosclerotic heart disease was brought to a hospital emergency room gasping for breath, cyanotic, and diaphoretic. She had had two myocardial infarctions in the previous 5 years, and during the year before the present episode she had been in stable chronic congestive heart failure. Over the 24 hours before being brought to the hospital she had become increasingly short of breath, and she had spent the night sitting up in a chair. She had taken extra diuretic tablets and nitroglycerine, but by midday she seemed to be failing rapidly and was rushed to the hospital by ambulance.

On arrival in the emergency room she was dyspneic and cyanotic, and her neck veins were distended. Her blood pressure was 70/30 mm Hg; her pulse was 140 per minute and irregular. There were moist rales throughout both lung fields. Examination of the heart revealed a prominent protodiastolic gallop, as well as a diffuse apical thrust in the anterior axillary line. There was 2+ pretibial and presacral edema.

While she was still in the emergency room, arterial and venous blood samples were obtained, an endotracheal tube was inserted, and she received assisted ventilation with supplemental oxygen. She was then transferred to the intensive care unit. Laboratory acid-base values are shown in Table 8-4.

Comment

With a low $[HCO_3^-]$ and a low pH, there is no doubt that the patient has a metabolic acidosis. One can see immediately, however, that a mixed disturbance must be present, since the PCO_2 is not decreased, as expected in uncomplicated metabolic acidosis (Table 3-2), but instead is slightly increased. Thus, there is also a superimposed respiratory acidosis — the acute form, due to pulmonary edema (see Chap. 6). The presence of two independently caused acidoses accounts for the extremely low pH in this patient.

Note that, viewed in isolation, the PCO_2 is just barely out of the upper range of normal (Tables 1-1 and 8-4); but when inter-

Table 8-4 : Results of laboratory tests in a 74-year-old woman
with pulmonary edema and shock
The numbers in parentheses are estimated values.

Test	On admission	4 Hours later	5 Days later
Arterial blood:			
pH	6.92	7.32	7.43
Pco_2 (mm Hg)	45	34	44
[H^+] (nmoles/L)	120 (120)	48 (48)	37 (37)
[HCO_3^-] (mmoles/L)	9 (9)	17 (17)	28 (29)
Po_2 (mm Hg)	35	100	110
Venous serum:			
[Na^+] (mmoles/L)	134	137	136
[K^+] (mmoles/L)	6.5	4.0	3.7
[Cl^-] (mmoles/L)	97	100	92
Total CO_2 (mmoles/L)	10	18	30
Anion gap (meq/L)	27	19	14

preted in light of a metabolic acidosis, it is distinctly abnormal. Even were the Pco_2 to fall below normal, one could still diagnose a superimposed respiratory acidosis if the Pco_2 was not lowered sufficiently. In the present example, we would anticipate a Pco_2 of 20 mm Hg if we were dealing with an uncomplicated metabolic acidosis (Table 3-3): For a fall in [HCO_3^-] of 15 mmoles per liter, the Pco_2 would be expected to fall by 1.3 × 15, or 20 mm Hg, for a new Pco_2 of 40 − 20, or 20 mm Hg. Thus, even if the Pco_2 had been 30 or 35 mm Hg — i.e., distinctly lower than normal — the diagnosis would still be that of a mixed metabolic and respiratory acidosis.

Identification of a primary respiratory component is absolutely critical to the proper management of this patient: It means (as described in Chap. 6, under Acute Pulmonary Edema, Comment) that immediate intubation of the trachea is required in order to lower the Pco_2 and promote oxygenation of the blood; therapy with HCO_3^- alone will not suffice.

A cause for the metabolic acidosis is suggested by the increased anion gap (Table 4-1). There was no history of ingestion of a toxin, renal function was only mildly impaired, as reflected in a serum creatinine concentration of 2.3 mg per 100 ml (204 μmoles/L; Table 1-1), and there were no detectable ketones in the urine or serum. The history and physical findings, with heart failure and shock, were most suggestive of lactic acidosis (Table 4-5), and this impression was borne out by a blood lactate level, reported subsequently, of 12 mmoles per liter.

The analysis should go one step further, to determine whether the elevation of the lactate can account for the entire metabolic acidosis. If one takes an average normal anion gap to be 12 meq per liter (Table 1-2), then the patient's gap has increased by 15 meq per liter (Table 8-4). This increment indicates that approximately 15 meq per liter of HCO_3^- was consumed in buffering a strong acid, and with a lactate concentration of 12 mmoles (or meq) per liter, one is able to conclude that virtually all the metabolic component can be accounted for by lactate. This conclusion is strengthened by the deduction that the patient had a relatively normal $[HCO_3^-]$ prior to her decompensation: If we add the 15 mmoles per liter of HCO_3^- consumed to the 9 mmoles per liter remaining, we obtain a normal value of 24 mmoles per liter. Although these estimates are only rough approximations, they can sometimes provide a clue to other causes of the metabolic acidosis (see the next clinical history, this chapter).

Treatment

As discussed in Chapter 4 (under Uncontrolled Diabetes Mellitus, Treatment, p. 92), metabolism of the circulating lactate, like that of other organic acids, is a potential source of new HCO_3^-, which will help to correct the metabolic acidosis if increased production of lactic acid can be halted. Because in this patient the increased production arose from hypotension and hypoxia (Table 4-5), the primary approach to treatment of the acid-base disorder was to raise Po_2 and lower Pco_2 by tracheal intubation and assisted ventilation with oxygen. In addition, the patient received 100 mmoles of $NaHCO_3$, as well as furosemide. As a result of this regimen, her blood pressure increased to 110/70 mm Hg, and the laboratory values improved rapidly (Table 8-4). The rise in $[HCO_3^-]$, from 9 to 17 mmoles per liter, is far greater than could be expected from the administered $NaHCO_3$ (see the above-cited section in Chap. 4). The simultaneous fall in the anion gap suggests that the source of the new HCO_3^- was the metabolism of endogenous lactate.

Subsequent laboratory tests (Table 8-4, 5 days later) showed virtually complete correction of all variables. Further evaluation of the heart disease disclosed no evidence of another myocardial infarction, and the acute decompensation was therefore attributed to progressive congestive heart failure and retention of fluid. Although the patient's acid-base disorders were readily corrected, her prognosis remains grave because of the underlying heart disease.

Vomiting and Hypotension

The patient is a 45-year-old man with a history of alcoholism characterized by binge drinking. The latest episode began approximately ten days before admission, when he stopped going out of his house and began drinking large amounts of beer,

**Table 8-5 : Results of laboratory tests in a 45-year-old man
with protracted vomiting and hypotension**
The numbers in parentheses indicate estimated values.

Test	On admission	2 Days later	1 Week later
Arterial blood:			
pH	7.32	7.48	—
P_{CO_2} (mm Hg)	28	46	—
[H^+] (nmoles/L)	48 (48)	33 (34)	—
[HCO_3^-] (mmoles/L)	14 (14)	33 (33)	—
Venous serum:			
[Na^+] (mmoles/L)	145	142	140
[K^+] (mmoles/L)	3.2	3.0	3.8
[Cl^-] (mmoles/L)	85	95	100
Total CO_2 (mmoles/L)	15	35	26
Anion gap (meq/L)	45	12	14

alone. As far as could be determined, he also stopped eating at about the same time. According to a sister who looked in on him, he began vomiting several days before admission, and the vomiting became worse until he was unable to keep down anything, even fluids. He steadfastly refused medical assistance and gradually became weaker. On the day of admission, his sister found him lying on the floor, unable to get up. He was then brought to the hospital by ambulance.

When the patient arrived in the emergency room, he was awake but disoriented and uncooperative. His temperature was 36°C, his blood pressure 85/40 mm Hg, pulse 110 per minute and regular, and respirations 24 per minute. His mucous membranes were very dry. There was no evidence of head trauma. The chest was clear, and examination of the heart was normal save for the tachycardia. The liver was firm, but neither tender nor enlarged. The spleen was also normal in size. There was no edema. Results of laboratory tests are shown in Table 8-5.

Comment

The arterial values indicate the presence of a metabolic acidosis, almost certainly due to lactic acidosis secondary to shock and tissue hypoxia (Table 4-5). The P_{CO_2} has fallen by 12 mm Hg, which is consistent with the anticipated response to uncomplicated metabolic acidosis (Table 3-3 and Fig. 3-2). Note, however, that the anion gap has risen more than would be expected for the observed reduction in [HCO_3^-]. In discussing the previous patient, we pointed out that one can estimate the amount of HCO_3^- (per liter of serum) that has been titrated by added acid when one subtracts the average normal value for the anion gap, namely, 12 meq per liter, from the measured anion gap. In the present instance, that calculation would be 45 − 12, or 33 meq per liter, which represents roughly the decrement in bicarbonate concentration that has been incurred by the metabolic acidosis.

If one adds this amount to the measured $[HCO_3^-]$ of 14 mmoles per liter (Table 8-5), the resultant value of 47 mmoles per liter indicates that the $[HCO_3^-]$ must have been very high prior to the development of the acidosis. Thus, the analysis to this point shows that a mixed disturbance must be present, and the clue to what that additional disorder is lies in the history and the venous values (Table 8-5).

On the basis of the prolonged vomiting, one would predict that the patient has a metabolic alkalosis (see Chap. 5), and the low serum $[Cl^-]$ (as well as the low $[K^+]$) supports this view (compare with Table 5-2). Metabolic acidosis alone does not lead to hypochloremia. Thus, this patient has a mixed disturbance of metabolic acidosis and metabolic alkalosis. Whether the pH falls within the acidemic range or the alkalemic range — or whether, conceivably, it is normal — will depend on which of the two disturbances predominates. When, as in the present instance, the pH is acid, some physicians would call this state a metabolic acidosis with the "footprints" of a metabolic alkalosis. We, however, prefer to call it a mixed disturbance of both disorders because the simultaneous presence of an alkalotic component has important therapeutic implications (discussed next).

Treatment

The sequence of events seen in this individual occurs commonly in binge drinkers. Pernicious vomiting leads to metabolic alkalosis, and then metabolic acidosis develops when volume contraction and shock set in. The importance of recognizing the alkalotic component, which is hidden, so to speak, by the acid pH, is that as the volume contraction and shock are corrected, metabolism of the circulating lactate will generate HCO_3^- endogenously, and severe metabolic alkalosis could result if exogenous HCO_3^- were administered as well. In this patient, treatment with isotonic saline (and some added K^+) corrected the shock and thereby, through generation of endogenous bicarbonate, raised the $[HCO_3^-]$ above normal and caused an alkalosis (Table 8-5, 2 Days later). Then gradually, with further repletion of the Cl^- deficit, the serum electrolytes returned to normal. The patient was discharged eight days after he had been admitted.

Acute Pulmonary Decompensation in Emphysema

The patient is a 73-year-old man with a long history of chronic obstructive lung disease and slowly declining pulmonary function. Until the present illness, he had had stable hypercapnia, with the Pco_2 ranging from approximately 52 to 58 mm Hg. Because of associated cor pulmonale and peripheral edema, he

**Table 8-6 : Results of laboratory tests in a 73-year-old man
with chronic obstructive pulmonary disease and acute pneumonia**
The numbers in parentheses indicate estimated values.

Test	On admission	2 Days later	1 Week later
Arterial blood:			
pH	7.33	7.42	7.41
P_{CO_2} (mm Hg)	78	61	52
$[H^+]$ (nmoles/L)	47 (47)	38 (38)	39 (39)
$[HCO_3^-]$ (mmoles/L)	40 (40)	38 (39)	32 (32)
P_{O_2} (mm Hg)	43	55	58
Venous serum:			
$[Na^+]$ (mmoles/L)	136	139	138
$[K^+]$ (mmoles/L)	3.6	3.9	4.2
$[Cl^-]$ (mmoles/L)	78	86	93
Total CO_2 (mmoles/L)	43	40	33
Anion gap (meq/L)	15	13	12

had been taking furosemide, 40 to 120 mg per day, for about six months. Over the five days prior to the present illness, he had developed increasing anorexia, malaise, and a cough productive of yellow sputum. He continued taking his medications, but finally was unable to keep them down. He became somnolent and disoriented, and therefore was brought to the hospital by his family.

Physical examination on admission showed the patient to be in obvious respiratory distress, with respirations at approximately 24 per minute and marked prolongation of the expiratory phase. He appeared drowsy, and when aroused was disoriented as to time and place. His rectal temperature was 38.5°C, pulse 110 per minute and regular, and blood pressure 110/70 mm Hg. There were scattered rhonchi and rales throughout both lung fields. The heart sounds were distant, but no gallops or murmurs were heard. The abdomen appeared normal. He had a trace of pitting edema in the ankles. Results of laboratory tests on the blood are shown in Table 8-6. An x-ray of the chest showed a new infiltrate in the right lower lobe, as well as evidence of long-standing emphysema.

Comment

Analysis of the arterial values is fully consistent with a chronic respiratory acidosis: With an increase in P_{CO_2} of 38 mm Hg (from 40 to 78), the expected rise in $[HCO_3^-]$ would be 0.4 × 38 (Table 3-3), or 15 mmoles per liter, for a new $[HCO_3^-]$ of 39 mmoles per liter. Plotting the arterial values on the confidence bands (Fig. 3-1) leads to the same conclusion. Yet we know from the history and physical examination that the patient had not only the chronic process (from emphysema) but also a superimposed acute respiratory acidosis (from pneumonia). In fact, the patient has not two but three acid-base disturbances, as will be demonstrated below. This patient thus illustrates a point emphasized in Chapter 3 (under Confidence Bands, p. 72), namely, that even when laboratory values clearly fall within a given confidence band, a mixed disturbance is not necessarily excluded. All relevant information must be considered, including the history and physical findings, in order to arrive at the correct diagnosis.

The full analysis of this patient's status proceeds as follows. We know that before the acute decompensation his P_{CO_2} had been in the 50s, with an average level of approximately 55 mm Hg. Thus, the P_{CO_2} of 78 mm Hg on admission (Table 8-6) almost certainly represented an acute hypercapnia of about 23 mm Hg (78 − 55) superimposed on a chronic hypercapnia of approximately 55 mm Hg. If only these two disorders were present, one would anticipate a $[HCO_3^-]$ in the low 30s: 15 × 0.4, or 6 mmoles per liter for the chronic hypercapnia, yielding a new $[HCO_3^-]$ of 24 + 6, or 30 mmoles per liter, and only a small additional rise for the superimposed acute hypercapnia. (Note: One cannot use the rules of thumb or confidence bands to assess the effect of an acute increase in P_{CO_2} when that increase begins at an abnormal, chronic level. These rules and bands are based on deviations in P_{CO_2} or $[HCO_3^-]$ from normal values.) However, we know that acute hypercapnia increases $[HCO_3^-]$ only minimally (by less than 1 to 3 mmoles per liter) in patients with chronic hypercapnia. Thus, the observed $[HCO_3^-]$ of 40 mmoles per liter (Table 8-6) is much too high for the acute hypercapnia superimposed on the chronic hypercapnia in this patient. This fact raises the suspicion that the patient has, in addition, a metabolic alkalosis.

The history of furosemide intake, coupled with a low serum $[K^+]$ and a low $[Cl^-]$ that is disproportionately reduced in relation to $[Na^+]$ (Table 8-6), all support the last diagnosis (see Chap. 5, under Administration of a Diuretic). Note that we can diagnose a component of metabolic alkalosis even though the pH is 7.33. Therefore, this patient appears to have a triple disturbance of acute respiratory acidosis, chronic respiratory acidosis, and metabolic alkalosis.

Treatment

Therapy with antibiotics and bronchodilators reversed much of the acute component of respiratory acidosis, and thereby uncovered the metabolic alkalosis more clearly; that is (Table 8-6, 2 Days later), the pH was no longer acid, and the $[HCO_3^-]$ of 38 mmoles per liter was higher than would be expected in a chronic respiratory acidosis with a P_{CO_2} of 61 mm Hg (Table 3-3). At the same time, the patient's alkalosis was treated with Cl^- as the K^+ salt (see Chap. 5, Administration of a Diuretic, Treatment). But because of the cor pulmonale, treatment with furosemide could not be stopped. Therefore, some component of metabolic alkalosis remained, as reflected in a $[HCO_3^-]$ slightly higher than the value predicted for this degree of chronic respiratory acidosis (Table 8-6, 1 Week later). The patient's prognosis is poor.

Vomiting, Hypotension, and Sepsis

A 38-year-old man who was receiving chemotherapy for widespread lymphoma developed persistent nausea and vomiting, which could not be controlled with antiemetics. One day before being seen by an oncologist, the patient developed fever and shaking chills. When examined the next morning, he appeared acutely and severely ill. His skin was warm and dry. His temperature was 38.5°C, his pulse 120 per minute and regular. Respirations were 30 per minute and deep but not labored, and blood pressure was 85/50 mm Hg. The lungs were clear to auscultation, and examination of the heart revealed the tachycardia but no murmurs, gallops, or rubs. There was no edema. As soon as blood had been drawn for laboratory tests (some results are shown in Table 8-7), treatment with intravenous fluids and broad-spectrum antibiotics was begun.

Comment

Viewed in isolation, the arterial blood gas studies are compatible with an uncomplicated acute respiratory alkalosis: The $[HCO_3^-]$ has fallen by 4 mmoles per liter, which is the predicted amount for a drop in P_{CO_2} of 20 mm Hg ($0.2 \times 20 = 4$; Table 3-3). Again, however, this isolated analysis is misleading, as could indeed be predicted by the history, which suggests that there should be a component of metabolic alkalosis due to the vomiting, and perhaps one of metabolic acidosis due to the hypotension. The fact

Table 8-7 : Results of laboratory tests in a 38-year-old man with vomiting, hypotension, and sepsis
The numbers in parentheses are estimated values.

Test	On evaluation	48 Hours later
Arterial blood:		
pH	7.62	7.44
P_{CO_2} (mm Hg)	20	35
$[H^+]$ (nmoles/L)	24 (25)	36 (36)
$[HCO_3^-]$ (mmoles/L)	20 (19)	23 (23)
P_{O_2} (mm Hg)	105	—
Venous serum:		
$[Na^+]$ (mmoles/L)	134	145
$[K^+]$ (mmoles/L)	2.5	3.2
$[Cl^-]$ (mmoles/L)	83	93
Total CO_2 (mmoles/L)	21	24
Anion gap (meq/L)	30	28

that the anion gap of 30 meq per liter is much higher than would occur in respiratory alkalosis, either acute or chronic (see discussion in conjunction with Table 3-4, p. 79), offers an immediate clue to the presence of a mixed disturbance.

Using the same approach as in the analysis of an earlier patient (Vomiting and Hypotension, this chapter), we can estimate that the component of metabolic acidosis (due to lactic acid; see below) reduced the $[HCO_3^-]$ by approximaely 30 − 12, or 18 meq per liter. This factor would have resulted in a $[HCO_3^-]$ of 24 − 18, or 6 mmoles per liter, were it not for the simultaneous presence of a metabolic alkalosis, which caused the measured $[HCO_3^-]$ to be 20 mmoles per liter. However, the two metabolic disorders alone — i.e., the acidosis and the alkalosis — cannot account for the degree of hypocapnia in this patient. Recall that secondary changes in Pco_2 in acid-base disorders are determined by changes of pH, which are sensed by central and peripheral chemoreceptors. Thus, the mild reduction in $[HCO_3^-]$ (and hence of pH) that resulted from the two metabolic acid-base disturbances would have induced only mild hyperventilation, reducing Pco_2 to approximately 35 mm Hg (40 − (1.3 × 4)), and yielding a pH of 7.38. The fact that the measured Pco_2 is much lower than the predicted value and that the pH is notably alkaline (Table 8-7) unmasks the acute respiratory alkalosis in this total analysis. Thus, the patient has a triple mixed disturbance of metabolic alkalosis, metabolic acidosis, and acute respiratory alkalosis. The associated hypokalemia and hypochloremia (Table 8-7) further support the diagnosis of the first of these components, i.e., of a metabolic alkalosis due to vomiting (see discussion in connection with Table 5-2).

Therefore, the chain of events in this patient was as follows: He had intractable vomiting for approximately four days, which led to metabolic alkalosis. He then contracted an infection with what turned out to be a gram-negative organism, and the resulting sepsis led to acute respiratory alkalosis and hypotension. The last two conditions — alkalemia and hypotension — stimulated production of lactic acid (see Chap. 4, under Lactic Acidosis, Treatment, p. 98), in this instance mainly by the lymphoma (Table 4-5), and thus brought about the third component of this triple disorder, metabolic acidosis.

Treatment

The coexistence of respiratory alkalosis and lactic acidosis complicates therapy because the first disorder aggravates the second. Bicarbonate could not be given to this patient, as it is ordinarily used for treatment of lactic acidosis, because the patient was already severely alkalemic (Table 8-7). Despite optimal therapy with antibiotics and isotonic saline, the large anion gap remained, indicating that production of lactic acid had not abated. Persistent lactic acidosis is known to occur in association with widespread lymphocytic tumors and leukemias (Table 4-5). It is an ominous prognostic sign. In this patient, the lactic acidosis continued until he died, several weeks later, from another episode of infection.

Selected References

General

Aberman, A., and Fulop, M. The metabolic and respiratory acidosis of acute pulmonary edema. *Ann. Intern. Med.* 76:173, 1972.

Anderson, R. J., Potts, D. E., Gabow, P. A., Rumack, B. H., and Schrier, R. W. Unrecognized adult salicylate intoxication. *Ann. Intern. Med.* 85:745, 1976.

Cohen, J. J., and Kassirer, J. P. *Acid-Base.* Boston: Little, Brown, 1982. Chap. 12.

Cohen, J. J., and Schwartz, W. B. Evaluation of acid-base equilibrium in pulmonary insufficiency. An approach to a diagnostic dilemma. *Am. J. Med.* 41:163, 1966.

Fulop, M., and Hoberman, H. D. Alcoholic ketosis. *Diabetes* 24:785, 1975.

Gabow, P. A., Anderson, R. J., Potts, D. E., and Schrier, R. W. Acid-base disturbances in the salicylate-intoxicated adult. *Arch. Intern. Med.* 138:1481, 1978.

Lyons, J. H., Jr., and Moore, F. D. Posttraumatic alkalosis: Incidence and pathophysiology of alkalosis in surgery. *Surgery* 60:93, 1966.

Narins, R. G., and Emmett, M. Simple and mixed acid-base disorders: A practical approach. *Medicine* 59:161, 1980.

Robin, E. D. Abnormalities of acid-base regulation in chronic pulmonary disease, with special reference to hypercapnia and extracellular alkalosis. *N. Engl. J. Med.* 268:917, 1963.

Answers to Problems

Problem 1-1
The data below were obtained on each of four patients.

Normal arterial values from Table 1-1: pH = 7.38 to 7.42; $[HCO_3^-]$ = 22 to 26 mmoles/L; P_{CO_2} = 37 to 43 mm Hg.

Cause of the disturbance	Arterial plasma			Type of disturbance
	pH	P_{CO_2} (mm Hg)	$[HCO_3^-]$ (mmoles/L)	
Prolonged vomiting	7.50	49	37	Metabolic alkalosis
Ingestion of NH_4Cl[a]	7.28	22	10	Metabolic acidosis
Anxiety-hyperventilation syndrome	7.57	24	21	Respiratory alkalosis (acute)[b]
Emphysema	7.33	68	35	Respiratory acidosis (chronic)[b]

[a] The net effect of ingesting NH_4Cl is the addition of hydrochloric acid, by the following reaction:

$$2\ NH_4Cl + CO_2 \rightarrow 2\ H^+ + 2\ Cl^- + H_2O + \underset{\text{urea}}{CO\ (NH_2)_2}$$

[b] For an explanation of the difference between acute and chronic respiratory disturbances, see the last paragraph under Compensatory Responses, Chapter 1, p. 12. Further discussions appear in Chapter 3 (under Confidence Bands and Rules of Thumb) and at the beginning of Chapter 6.

Problem 1-2

Respiratory Acidosis. When CO_2 is retained, as during alveolar hypoventilation due to barbiturate intoxication, the P_{CO_2} rises. As the CO_2 is hydrated (Eq. 1-1) or hydroxylated (Eq. 1-2), H^+ and HCO_3^- are produced; this H^+ must be buffered by nonbicarbonate buffers (mainly hemoglobin, other proteins, and organic phosphates; Fig. 1-4), and in the process HCO_3^- rises (Fig. 1-7). Since the primary disturbance is respiratory, the compensatory response will be renal, and it takes several days to come to completion. The compensation involves increased reabsorption of HCO_3^- occasioned by the elevated P_{CO_2} (Fig. 2-2B).

Note that here is an instance in which the plasma HCO_3^- concentration rises during acidosis. Inasmuch as HCO_3^- is a major buffer, there is a reflex tendency for the neophyte in acid-base balance to predict a decrease in plasma HCO_3^- during acidosis. The key to understanding why it increases during respiratory acidosis (but decreases during metabolic acidosis) is to remember that the H^+ derived from the processing of CO_2 (Eqs. 1-1 and 1-2) cannot be buffered by the HCO_3^-/CO_2 system.

Metabolic Alkalosis. A net loss of H^+ from the body, as occurs through the loss of gastric HCl during prolonged vomiting, is accompanied by an increase in the plasma HCO_3^- concentration. This fact is evident from Equation 1-11, which will be driven to the left as HCl is withdrawn. The respiratory compensation for this primary metabolic disturbance is a diminution in alveolar ventilation due to the alkalosis, and a consequent rise in P_{CO_2}.

Respiratory Alkalosis. The primary event in this disturbance of H^+ balance is alveolar hyperventilation and a decline in P_{CO_2}. This change will drive the reactions shown at the top of Figure 1-7 to the left, and the plasma HCO_3^- concentration therefore will decrease. The compensatory response by the kidneys — which will begin within minutes of the onset of the disturbance but will take days to take full effect — is to decrease the reabsorption of HCO_3^- as a consequence of the lowered P_{CO_2} (Fig. 2-2B).

Arterial pH in Mixed Disturbances. The reason for this question is to point out that the plasma pH can be predicted only if the two components shift the pH in the same direction. Thus, a mixture of two acidoses will certainly result in an acidotic pH, and two alkaloses in an alkalotic pH. When, however, the two components shift the pH in opposite directions (i.e., a mixture of an acidosis and an alkalosis), the plasma pH might be alkalotic, normal, or acidotic, depending on which component predominates.

Problem 2-1

The Henderson-Hasselbalch equation, being an expression of the ionization or dissociation properties of acids and bases, can be utilized to solve this problem.

$$pH = pK' + \log \frac{[\text{base; i.e., } H^+ \text{ acceptor}]}{[\text{acid; i.e., } H^+ \text{ donor}]}$$

For phenobarbital,

$$pH = 7.2 + \log \frac{[\text{Ionized form}]}{[\text{Nonionized form}]}$$

Plasma, pH 7.3

$$7.3 = 7.2 + \log \frac{[\text{Ionized form}]}{[\text{Nonionized form}]}$$

$$0.1 = \log \frac{[\text{Ionized form}]}{[\text{Nonionized form}]}$$

$$\therefore \frac{[\text{Ionized form}]}{[\text{Nonionized form}]} = \frac{1.26}{1}$$

The concentration of total unbound phenobarbital is 6.0 mg/ 100 ml plasma; $\frac{1.26}{2.26}$ of this total exists in the ionized form, and $\frac{1.00}{2.26}$ of the total exists in the nonionized form. Hence:

$$[\text{Ionized form}] = \frac{1.26}{2.26} \times 6.0 = 3.3 \text{ mg/100 ml}$$

$$[\text{Nonionized form}] = \frac{1.00}{2.26} \times 6.0 = 2.7 \text{ mg/100 ml}$$

Plasma, pH 7.7

$$7.7 = 7.2 + \log \frac{[\text{Ionized form}]}{[\text{Nonionized form}]}$$

$$0.5 = \log \frac{[\text{Ionized form}]}{[\text{Nonionized form}]}$$

$$\therefore \frac{[\text{Ionized form}]}{[\text{Nonionized form}]} = \frac{3.16}{1}$$

$$\therefore [\text{Ionized form}] = \frac{3.16}{4.16} \times 6.0 = 4.6 \text{ mg/100 ml plasma}$$

$$[\text{Nonionized form}] = \frac{1.00}{4.16} \times 6.0 = 1.4 \text{ mg/100 ml plasma}$$

Urine, pH 5.2

$$5.2 = 7.2 + \log \frac{\text{[Ionized form]}}{\text{[Nonionized form]}}$$

$$-2.0 = \log \frac{\text{[Ionized form]}}{\text{[Nonionized form]}}$$

$$2.0 = \log \frac{\text{[Nonionized form]}}{\text{[Ionized form]}}$$

$$\therefore \frac{\text{[Nonionized form]}}{\text{[Ionized form]}} = \frac{100}{1}$$

i.e., when the reaction of the urine is acid, most of the phenobarbital exists in the nonionized form, which can diffuse across the membranes of tubular cells and hence can be passively reabsorbed.

Urine, pH 8.2

$$8.2 = 7.2 + \log \frac{\text{[Ionized form]}}{\text{[Nonionized form]}}$$

$$1.0 = \log \frac{\text{[Ionized form]}}{\text{[Nonionized form]}}$$

$$\therefore \frac{\text{[Ionized form]}}{\text{[Nonionized form]}} = \frac{10}{1}$$

i.e., when the reaction of the urine is alkaline, most of the phenobarbital exists in the ionized form, to which renal tubular cells are relatively impermeable. Hence, alkalinization of the urine can markedly diminish the reabsorption and thus enhance the renal excretion of a weak acid, such as phenobarbital.

	Total unbound phenobarbital in plasma (mg/100 ml)	Ratio of unbound phenobarbital: [Ionized] / [Nonionized]	Plasma concentration of unbound phenobarbital	
			Ionized	Nonionized (mg/100 ml)
Plasma, pH 7.3	6.0	$\frac{1.26}{1}$	3.3	2.7
Plasma, pH 7.7	6.0	$\frac{3.16}{1}$	4.6	1.4
Urine, pH 5.2	—	$\frac{1}{100}$	—	—
Urine, pH 8.2	—	$\frac{10}{1}$	—	—

Note that alkalinization may have a further advantage. Giving $NaHCO_3$ alkalinizes not only the urine but also the plasma (Eq. 1-9). This change reduces the concentration of the nonionized form in plasma. Since this is the form that passes most readily across cell membranes, including those of the brain, alkalinization probably reduces the concentration of phenobarbital in cerebral cells, and thereby hastens recovery from coma.

The beneficial effects of increased urine flow and alkalinization during experimental phenobarbital intoxication were presented in the following paper: Waddell, W. J., and Butler, T. C. The distribution and excretion of phenobarbital. *J. Clin. Invest.* 36:1217, 1957. Two clinical examples in which the principle of nonionic diffusion was applied to salicylate therapy and intoxication appear in: Levy, G., et al., *N. Engl. J. Med.* 293:323, 1975; and Hill, J. B., *N. Engl. J. Med.* 288:1110, 1973 (see also Chap. 8, under Overdose with Salicylate, Treatment).

Problem 2-2

It is clear from Figure 2-A that, because the NH_3/NH_4^+ buffer system has a pK' of 9.2, NH_4Cl cannot be a titratable acid (T.A.). Recall that T.A. is defined as the amount of strong base that must be added to acid urine in order to return the pH of that urine to 7.40. If urine is thus titrated from a minimal pH of 4.4

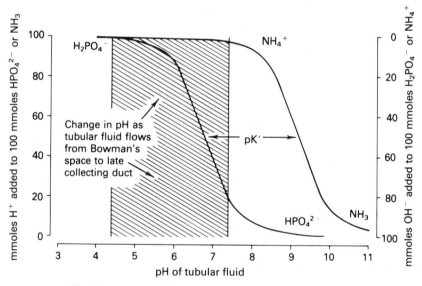

Fig. 2-A
Behavior of the buffer systems, $HPO_4^{2-}/H_2PO_4^-$ and NH_3/NH_4^+, in tubular fluid. Note that as OH^- is added to acid urine (ordinate on right) most of the phosphate is converted to the HPO_4^{2-} form, whereas the vast majority of the ammonia system remains in the NH_4^+ form. In other words, NH_4^+ salts cannot be appreciably titrated from an acid pH to 7.4, and therefore they are not titratable acids.

(the shaded area in Fig. 2-A), very little NH_4^+ will be converted to NH_3 — a consequence of the fact that a buffer is most effective within ± 1 pH unit of its pK'.

Problem 2-3

Ordinarily, virtually all the filtered HCO_3^- is reabsorbed (see Reabsorption of Filtered HCO_3^-, p. 37). Suppose that the urinary concentration of HCO_3^- is 1.0 mmole per liter. Then, applying this value and a P_{CO_2} for urine of 40 mm Hg to the Henderson-Hasselbalch equation (Eq. 1-9) yields a urinary pH of approximately 6; and if the urinary HCO_3^- concentration is reduced to 0.1 mmole per liter, the urinary pH will be approximately 5. These calculations show that the urinary pH could be lowered almost to the minimal value merely by reabsorbing virtually all the filtered HCO_3^-. Even though that happens normally, the situation is nevertheless theoretical because titratable acids (mainly NaH_2PO_4) are formed simultaneously, as shown in Figure 2-8.

Problem 3-1

1. Both the calculation and the method of estimation used in this example are described in the section Conversion of pH to $[H^+]$, on page 63.
2. Calculation of P_{CO_2} from the Henderson-Hasselbalch equation (Eq. 3-6):

$$pH = 6.10 + \log \frac{[HCO_3^-]}{0.03 \times P_{CO_2}}$$

$$7.32 = 6.10 + \log \frac{15}{0.03 \times P_{CO_2}}$$

$$1.22 = \log \frac{15}{0.03 \times P_{CO_2}}$$

$$\frac{15}{0.03 \times P_{CO_2}} = \text{antilog } 1.22$$

$$\frac{15}{0.03 \times P_{CO_2}} = 16.60$$

$$P_{CO_2} = \frac{15}{16.60 \times 0.03}$$

$$P_{CO_2} = 30.1 \text{ mm Hg}$$

Estimation of P_{CO_2} from the Henderson equation (Eq. 3-4):

$$[H^+] = 24 \frac{P_{CO_2}}{[HCO_3^-]}$$

By estimation, the $[H^+]$ corresponding to a pH of 7.32 is 48 nmoles/L (40 + 8). Therefore:

$$48 = 24\,\frac{P_{CO_2}}{15}$$

$$P_{CO_2} = \frac{15 \times 48}{24}$$

$$P_{CO_2} = 30 \text{ mm Hg}$$

3. Calculation of $[HCO_3^-]$ from the Henderson-Hasselbalch equation:

$$pH = 6.10 + \log \frac{[HCO_3^-]}{0.03 \times P_{CO_2}}$$

$$\frac{[HCO_3^-]}{0.03 \times 44} = \text{antilog } (7.55 - 6.10)$$

$$[HCO_3^-] = 0.03 \times 44 \times 28.18$$

$$[HCO_3^-] = 37.2 \text{ mmoles/L}$$

Estimation of $[HCO_3^-]$ from the Henderson equation:

$$[H^+] = 24\,\frac{P_{CO_2}}{[HCO_3^-]}$$

First, estimate $[H^+]$ corresponding to a pH of 7.55 (see "80 rule," p. 65, and Table 3-1): for pH 7.50, estimated $[H^+] = 0.8 \times 40$, or 32 nmoles/L; for pH 7.60, estimated $[H^+] = 0.8 \times 32$, or 26 nmoles/L. By interpolation, estimated $[H^+]$ for pH 7.55 $= 26 + [0.5 \times (32 - 26)]$, or 29 nmoles/L. Then, substituting in Henderson equation:

$$29 = 24\,\frac{44}{[HCO_3^-]}$$

$$[HCO_3^-] = \frac{24 \times 44}{29}$$

$$[HCO_3^-] = 36 \text{ mmoles/L}$$

4. Calculation of pH from $[H^+]$ (Eq. 3-2):

$$[H^+] = \text{antilog } (9 - pH)$$

$$pH = 9 - \log [H^+]$$

$$pH = 9 - \log 65$$

$$pH = 7.187$$

Table 3-A : Tabulated solutions to questions asked in Problem 3-1

	pH		$[H^+]$ (nmoles/L)		P_{CO_2} (mm Hg)		$[HCO_3^-]$ (mmoles/L)	
	Calc.	Est.	Calc.	Est.	Calc.	Est.	Calc.	Est.
(1)	7.40	—	40	40	40	—	24	—
(2)	7.32	—	—	—	30	30	15	—
(3)	7.55	—	—	—	44	—	37	36
(4)	7.19	7.19	65	—	75	—	28	—

Estimation of pH from $[H^+]$, using the "80 rule" (Table 3-1): For pH 7.10, estimated $[H^+] = 0.8 \times 100$, or 80 nmoles/L; for pH 7.20, estimated $[H^+] = 0.8 \times 80$, or 64 nmoles/L. By interpolation,

$$pH = 7.20 - [0.1 \div (80 - 64)]$$

$$pH = 7.194$$

The values and units that were requested in Problem 3-1 are shown in italics in Table 3-A.

Problem 3-2

The history of vomiting when the patient was at the other hospital, in conjunction with the volume contraction that the patient showed on physical examination and the report of a high plasma $[HCO_3^-]$, raised the suspicion of a metabolic alkalosis. When the values on admission, which are listed in Table 3-B, were plotted on Figure 3-2, it turned out that the point fell within the confidence band in Figure 3-2A, but below the band in Figure 3-2B. This discrepancy, which is also apparent if the values are plotted on Figure 3-1A and B, is impossible, because both parts of Figures 3-1 and 3-2 are based on the same mathematical relationship, the Henderson-Hasselbalch equation. The physician thus knew immediately that a laboratory error had been made. (Even allowing the $[HCO_3^-]$ in *venous* plasma to be 2 to 3 mmoles per liter higher than in arterial plasma would not eliminate the discrepancy.)

Checking the Henderson-Hasselbalch relationship by arithmetic approximation confirmed the presence of a laboratory error. Using the reported pH of 7.51 and a $[HCO_3^-]$ of 42 mmoles per liter (as the approximate arterial value), the arterial P_{CO_2} was estimated to be 54 mm Hg. First, pH 7.51 corresponds approximately to a $[H^+]$ of 31 nmoles per liter; then, according to Equation 3-4,

$$[H^+] = 24 \frac{P_{CO_2}}{[HCO_3^-]}$$

$$31 = 24 \frac{P_{CO_2}}{42}$$

$$P_{CO_2} = \frac{42 \times 31}{24}$$

$$P_{CO_2} = 54.3 \text{ mm Hg}$$

Alternatively, by using the reported pH of 7.51 and P_{CO_2} of 34 mm Hg, the $[HCO_3^-]$ for arterial plasma was estimated from the above equation to be 26 mmoles per liter:

$$[HCO_3^-] = \frac{24 \times 34}{31}$$

$$[HCO_3^-] = 26.3 \text{ mmoles per liter}$$

The physician therefore asked for repeat laboratory determinations, the results of which are listed in the final column of Table 3-B. In the meantime, however, he had already concluded on the basis of the following reasons that the patient almost certainly had a metabolic alkalosis, and that the P_{CO_2} was therefore more likely to be the erroneous value: (1) There was a history of vomiting, and (2) a high $[HCO_3^-]$ had been determined at the other hospital. His deduction was borne out by the repeat measurements (Table 3-B).

Armed with this new information, the physician sought a cause for the metabolic alkalosis. It had been suspected that the patient might have the so-called milk-alkali syndrome, because of the prior history of unexplained hypercalcemia, a BUN/creatinine ratio that had repeatedly far exceeded the usual value of approximately 10, hence suggesting a very high protein intake, and a

Table 3-B : Results of laboratory tests in a 61-year-old woman immediately on admission to the hospital and three hours later

Test	On admission	Three hours after admission
Arterial blood:		
pH	7.51	7.48
P_{CO_2} (mm Hg)	34	55
Venous serum:		
$[Na^+]$ (mmoles/L)	152	—
$[K^+]$ (mmoles/L)	3.6	—
$[Cl^-]$ (mmoles/L)	98	—
Total CO_2 (mmoles/L)	45	42
BUN (mg/100 ml)	82	—
Creatinine (mg/100 ml)	3.1	—

history of a "strange personality" and possible food fads. On further questioning it was learned that the patient had indeed for years drunk 4 to 5 liters of milk daily. When this pattern was ended, she recovered.

It should be emphasized that this is a true history, and that laboratory errors do occur but, by considering all the values in the light of the patient's history and physical findings, one can nevertheless reach a correct diagnosis. The chronic renal failure in this patient may have resulted from chronic infection, as had been suspected, or it may have been due to the milk-alkali syndrome. In either case, patients who ingest large amounts of alkali are more likely to develop severe metabolic alkalosis if they have concurrent renal failure (Table 5-1).

Index